INVERTEBRATE LEARNING

Volume 3
Cephalopods and Echinoderms

INVERTEBRATE LEARNING

Volume 1 • Protozoans Through Annelids
Volume 2 • Arthropods and Gastropod Mollusks
Volume 3 • Cephalopods and Echinoderms

INVERTEBRATE LEARNING

Volume 3
Cephalopods and Echinoderms

Edited by
W. C. Corning and J. A. Dyal
Department of Psychology
University of Waterloo
Waterloo, Ontario, Canada

and

A. O. D. Willows
Department of Zoology
University of Washington
Seattle, Washington

PLENUM PRESS · NEW YORK AND LONDON

Library of Congress Cataloging in Publication Data

Corning, William C
Invertebrate learning.

Includes bibliographies.
CONTENTS: v. 1. Protozoans through Annelids.–v. 2. Arthropods and Gastropod Mollusks.–v. 3. Cephalopods and Echinoderms.
1. Invertebrates–Psychology. 2. Invertebrates–Behavior. I. Dyal, James Albert, 1928- joint author. II. Willows, A. O. D., joint author. III Title.
QL364.2.C67 156'.3'15 72-90335
ISBN 0-306-37673-3 (v. 3)

A Division of Plenum Publishing Corporation
227 West 17th Street, New York, N.Y. 10011

United Kingdom edition published by Plenum Press, London
A Division of Plenum Publishing Company, Ltd.
4a Lower John Street, London W1R 3PD, England

Printed in the United States of America

CONTENTS OF VOLUME 3

Contents of Volume 1 ix

Contents of Volume 2 xi

Chapter 11

The Cephalopods **1**

G. D. Sanders

I. Introduction 1
II. Behavioral Research with Cephalopods 2
A. Some Advantages of Working with Cephalopods ... 2
B. The Disadvantages of Working with Cephalopods ... 2
C. A Guide to Recent Review Literature 3
III. Cephalopod Biology 3
A. Classification 3
B. Evolution 5
C. Morphology, Physiology, and Reproduction 6
D. The Major Sense Organs: The Statocysts, Suckers, and Eyes 9
E. The Nervous System 10
IV. Early Studies of Learning 13
A. Avoidance of Sea Anemones 13
B. Habituation 14
C. Conditioning 14
D. Discrimination Learning 15
E. Detour Experiments 15
F. Problem Boxes 17
G. The Early Learning Experiments of Boycott and Young 17

V. Types of Learning 19
A. Habituation 19
B. Conditioning 20
C. Associative Learning 20
D. Visual Discrimination 25
E. Discrimination by Touch 33
VI. Other Learning Phenomena 35
A. Proprioception and Learning 35
B. Reversal Learning 36
C. Generalization and Transfer 40
D. Delayed Response and Delayed Reinforcement 43
E. Intertrial Interval and Learning 47
F. Retention of Learned Behaviors 49
VII. Brain Lesions and Learning 54
A. Separate Centers for Visual and Tactile Learning 54
B. The Visual System 56
C. The Tactile System 63
D. The Vertical Lobe System and Learning in the Cuttlefish 68
E. The Vertical Lobe and Visual Discrimination Learning in Octopuses 70
F. The Vertical Lobe and Tactile Discrimination Learning in Octopuses 77
G. Delayed Response, Delayed Reinforcement, and the Vertical Lobe 82
H. The Effect of Vertical Lobe Removal: A Summary 84
I. Proposed Functions of the Vertical Lobe System 86
VIII. Electrophysiological Recording from the Cephalopod Nervous System 90
IX. Conclusions 92
References 93

Chapter 12

The Echinoderms **103**

A. O. D. Willows and W. C. Corning

I. Introduction 103
II. Taxonomy and Evolution 106
A. Classification 106
B. Evolutionary Relationships 108

III. Sensory-Motor Apparatus in Echinoderms 109
A. Organization of the Nervous System 109
B. The Ectoneural, Hyponeural, and Entoneural Systems 113
C. General Behavior 114
IV. Learning Demonstrations 118
A. Righting Reflex and Persistence of Movement 119
B. Escape Behavior 122
C. Associative Learning 126
V. Conclusions 134
References 134

Editor's Note **137**

W. C. Corning, J. A. Dyal, and A. O. D. Willows

Chapter 13

Critical Commentary **139**

M. E. Bitterman

Chapter 14

Synthesis: A Comparative Look at Vertebrates **147**

R. Lahue and W. C. Corning

I. Examples of Functional Convergence 148
A. Equipotentiality 148
B. Dominant Focus 149
C. Incremental and Decremental Processes 151
D. Memory Transfer 153
E. Unit Conditioning 153
F. Image-Driven Behavior 154
G. Unit Specifications 154
H. Biochemical Correlates 155
II. Divergent Properties 156
A. Cell Morphology 156
B. Specification of Cell Type, Structure, and Location 157
C. Nerve Process Differentiation 157
D. Feedback Circuitry 157

E. Glial Presence ... 158
F. The Numbers of Neurons and Their General Organization ... 158
G. Integration/Information Processing ... 158
III. Some Comparative Strategies and Problems ... 159
IV. The Evolution of Learning Strategies in Invertebrates and Vertebrates: Some Speculation ... 168
A. Preprogrammed Morphological/Physiological Strategies ... 168
B. Epigenetic Strategies ... 168
C. Acquired Relations Between System Components (Suprastructural) ... 169
References ... 173

Appendix A
Learning in Bacteria, Fungi, and Plants ... 179
Philip B. Applewhite

Appendix B
Progressions: Bibliography of Recent Publications ... 187
Sonja Ziganow and W. C. Corning

Index ... 213

CONTENTS OF VOLUME 1

Chapter 1
Invertebrate Learning and Behavior Taxonomies
J. A. Dyal and W. C. Corning

Chapter 2
Protozoa
W. C. Corning and R. Von Borg

Chapter 3
Behavioral Modifications in Coelenterates
N. D. Rushforth

Chapter 4
Platyhelminthes: The Turbellarians
W. C. Corning and S. Kelly

Chapter 5
Behavior Modifications in Annelids
J. A. Dyal

CONTENTS OF VOLUME 2

Chapter 6

The Chelicerates
Robert Lahue

Chapter 7
Learning in Crustacea
Franklin B. Krasne

Chapter 8
Learning in Insects Except Apoidea
Thomas M. Alloway

Chapter 9
Honey Bees
Patrick H. Wells

Chapter 10
Learning in Gastropod Mollusks
A. O. D. Willows

Chapter 11

THE CEPHALOPODS

G. D. SANDERS

Psychology Department
City of London Polytechnic
London E1 7PF, England

I. INTRODUCTION

The cephalopods are marine predators that have survived by successfully competing with the teleost fish in all areas of the sea from shoreline to abyssal depths. The brain of cephalopods approaches that of the vertebrates in relative size, being as large as or larger than the brains of many fish although smaller than those of birds and mammals. The extensive development of sensory and neural systems that is seen in the recent cephalopods enables them to exhibit the most complex types of adaptive behavior that are found in the nonvertebrate world. Having evolved as a separate group for more than 500 million years to reach the position that they now occupy in the animal kingdom, cephalopods are of obvious interest to investigators employing a comparative approach. It is unfortunate that the practical difficulties involved in obtaining, transporting, and housing cephalopods in conditions suitable for behavioral research have precluded their wider use.

The choice of experimental animals from among the cephalopods has, to a large extent, been governed by their adaptability to laboratory conditions. The common squid, *Loligo vulgaris*, is easily disturbed and readily sustains injuries to the posterior end of its mantle. Consequently, this cephalopod is difficult to keep for long periods in captivity, although some success has been achieved by the use of a large, circular island tank. The cuttlefish, *Sepia officinalis*, is prone to similar damage but has been successfully maintained for several weeks in the laboratory. *Octopus vulgaris* adapts readily to life in aquaria and the vast majority of behavioral studies have employed this species. A few investigators have used cuttlefish, while squids have not been used at all for studies of learning, although their giant nerve fibers have been widely utilized by neurophysiologists for the study of neural transmission. In this chapter the terms octopus and octopuses refer to *O. vulgaris*, while other species of octopods are referred to by their generic and specific names. "Cuttlefish" is synonymous with *S. officinalis* and squid refers to *L. vulgaris* unless otherwise stated.

II. BEHAVIORAL RESEARCH WITH CEPHALOPODS

A. Some Advantages of Working with Cephalopods

Cephalopods are readily available in those coastal regions where they are habitually fished and eaten (e.g., the Mediterranean and Japan). *O. vulgaris* adapts rapidly to life in the laboratory and specimens have been kept successfully for more than six months. Octopuses are voracious feeders and are not easily satiated so that motivation is rarely a problem, although they become rather sluggish at temperatures below 12°C. Their visual and tactile receptor systems are exceptionally well developed, and to a large extent the associated neural centers are separated within the brain. Octopuses are particularly suitable for lesion studies as they withstand considerable surgical interference and recover rapidly. Small lesions are closed by contraction of the musculature and no special aseptic operating conditions are required (Young, 1971, Appendix I). The higher centers of the brain are readily accessible through the soft cartilaginous cranium, in which the brain is enclosed. As an added advantage large areas of the upper part of the brain (the supraesophageal lobes) may be removed without interference to the sensory-input and motor-response systems. This is particularly true of the tactile centers where sensory input comes from below and the motor responses are organized within the nervous system of the arms. The ability of octopuses to master a variety of visual, tactile, and chemotactile discrimination tasks makes them suitable subjects for behavioral studies.

B. The Disadvantages of Working with Cephalopods

The main obstacles to a wider use of cephalopods are the practical difficulties of transportation and housing. Small *vulgaris* of 100–200 g have been successfully sent by air freight from Naples to London. When housed in plastic bags with a small volume of water, just sufficient to cover the animal, and about 8 liters of oxygen the octopuses survive well for up to 10 hr. Prescott and Brosseau (1962) have described a refrigeration technique for the shipment of the giant Pacific octopus (*O. appollyon*) that may prove adaptable for use with other species.

Although cephalopods are available at many of the world's marine stations, a considerable installation of equipment is required to house sufficient numbers for behavioral studies. To date the animals used in experiments come directly from the sea, allowing the investigator no control over, nor precise knowledge of, their previous experience. This problem is, however, technically solvable now that *Octopus vulgaris* (Itami *et al.*, 1963), *Octopus briareus* (Wolterding, 1971), *Octopus joubini* (Boletzky and Boletzky, 1969; Bradley, 1974; Thomas and Opresko, 1973), and cuttlefish and squid (Ohshima and Sang Choe, 1961, 1963) have been bred in the laboratory. *O. joubini* and *O. briareus* may prove to be of particular interest as the young do not pass through the planktonic larval stage, like *O. vulgaris*, but begin their life on the bottom as small versions of the adult immediately upon hatching.

C. A Guide to Recent Review Literature

An interesting account of cephalopods may be found in Packard's (1972) discussion of convergent evolution between cephalopods and fish. Young (1971) has provided a monumental coverage of the nervous system of *O. vulgaris*. Boycott (1954), Mackintosh (1965*a*), Wells (1962*a*, 1965*a,b,c*, 1966), and Young (1961*a*, 1964*a*) have reviewed behavioral experiments with octopuses. The paper by Boycott also describes methods of fishing and keeping octopuses. More popular accounts of cephalopods are those of Lane (1957) and Cousteau and Diole (1973). The earlier work contains a good general bibliography while the latter is well illustrated, with numerous photographs in color.

III. CEPHALOPOD BIOLOGY

A. Classification

The cephalopods are marine invertebrates that together form a class of the phylum Mollusca. Classification has been hampered by the fact that the majority of the known species are extinct. They were divided into two subclasses, the Tetrabranchiates (Ectocochlia) and the Dibranchiates (Coleoidea), but recently the following classification has become generally accepted:

CLASS CEPHALOPODS

SUBCLASS NAUTILOIDEA. . Cephalopods with coiled or straight external shells divided into chambers by septa with simple sutures. All of the members are now extinct except for one surviving genus, *Nautilus* (the pearly nautilus).

SUBCLASS AMMONOIDEA. . Cephalopods with coiled external shells divided into chambers by septa with complex sutures. They are all extinct.

SUBCLASS COLEOIDEA. . . . . Cephalopods with an internal shell considerably reduced or without a shell.

Order Belemnoidea. . . . . Squidlike coleoids with a straight internal shell. They flourished during the Mesozoic era and became extinct at the end of the Cretaceous period.

Order Vampyromorpha. . . . . . The most primitive living order of coleoids, represented by a single species, *Vampyroteuthis infernalis*, a deep bathypelagic cephalopod.

Order Sepioidea. . . . . Coleoids with eight arms and a pair of tentacles that are modified for catching prey and can be retracted into pockets beneath the eyes, e.g., the cuttlefish, *S. officinalis*.

Order Teuthoidea. . . . . Coleoids with eight arms and a pair of tentacles that are never retracted into special pockets. The majority of living cephalopods belong to this order.

Suborder Myopsida. . . . . . . Teuthoids with the cornea almost completely closed over the eye, e.g., the common squid, *L. vulgaris*.

Suborder Oegopsida. . . . . . . Teuthoids in which the cornea is not closed over the eye, e.g., many oceanic squids including the giant squids belonging to the genus *Architeuthis*.

Order Octopoda. . . . . Coleoids with eight arms, a rounded body, and, usually, only a minute vestige of the shell. The order includes the benthic octopods, *O. vulgaris* (the common octopus) and *Eledone moschata* (the lesser octopus), as well as the pelagic paper nautilus (*Argonauta*).

Among the living cephalopods the Sepioidea and Teuthoidea with eight arms and two tentacles are often referred to as decapods to distinguish them from the eight-armed octopods.

B. Evolution

Although the majority of cephalopods are now extinct the coleoids are probably only just reaching their evolutionary peak. The earliest known cephalopod species are nautiloids from the Upper Cambrian. Nautiloids flourished throughout the Paleozoic era but today only the "living fossil" *Nautilus* survives. The ammonoid and belemnoid cephalopods arose alongside the nautiloids and flourished during the Mesozoic, only to die out completely in the Cretaceous. The coleoids appeared at the end of the Paleozoic and subsequently gave rise to all recent forms.

Cephalopod evolution has been characterized by a progressive loss of the protective shell, presumably accompanied by a change in the behavior of the majority of living species from passive defense to the role of active predators. This change in behavior has depended upon an improvement of the sensory and nervous systems with a switch in emphasis from chemotactile to visual input and the development of the capacity to learn. The centers associated with vision and learning that have been identified in

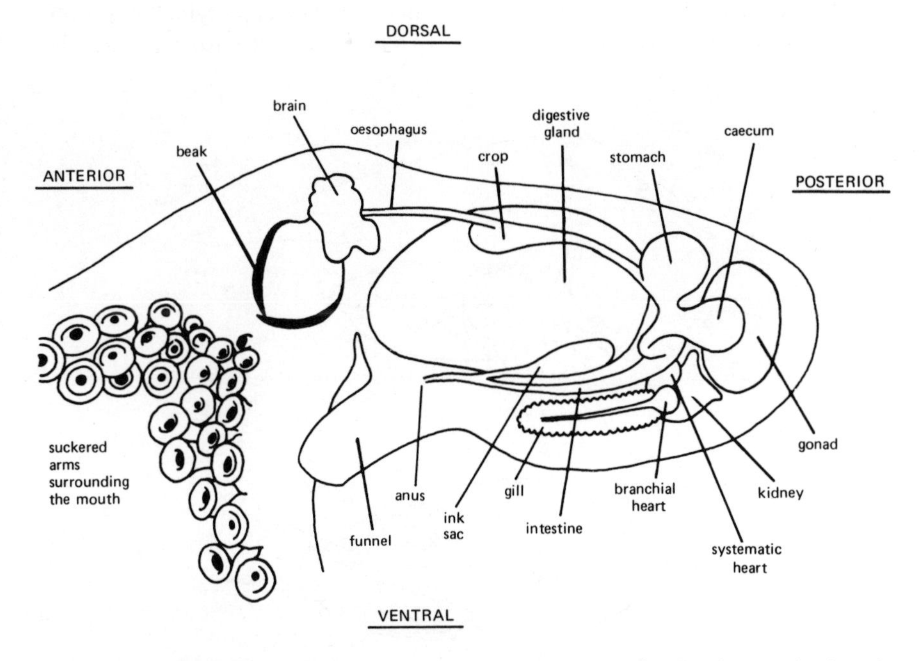

Fig. 1. A sagittal section of the entire body of a young octopus showing the organization of its main features.

octopus are poorly developed in the primitive *Nautilus*. As a group, the modern coleoids evolved in direct competition with the teleost fish, and many instances of convergent evolution may be traced (Packard, 1972).

In cephalopods the molluscan visceral hump has become elongated and the functional orientation of the body has changed. The original ventral surface has become the functional anterior end, while the posterior end is the original dorsal surface of the visceral hump. The funnel, morphologically posterior, is situated on the functionally ventral surface (Fig. 1). It is now customary to use the functional axes rather than the original morphological ones.

C. Morphology, Physiology, and Reproduction

A few of the commonly known cephalopods are illustrated in Fig. 2. The body consists of three main parts: the arms, the head, and the mantle that surrounds the visceral mass (Fig. 1). The head bears a pair of large eyes and a ring of arms around the mouth. Octopuses have eight arms with two rows of suckers on the inner surface extending from the mouth to the arm tips. The third right arm of the male, the hectocotylus, has been modified for passing spermatophores (packets of spermatozoa) into the

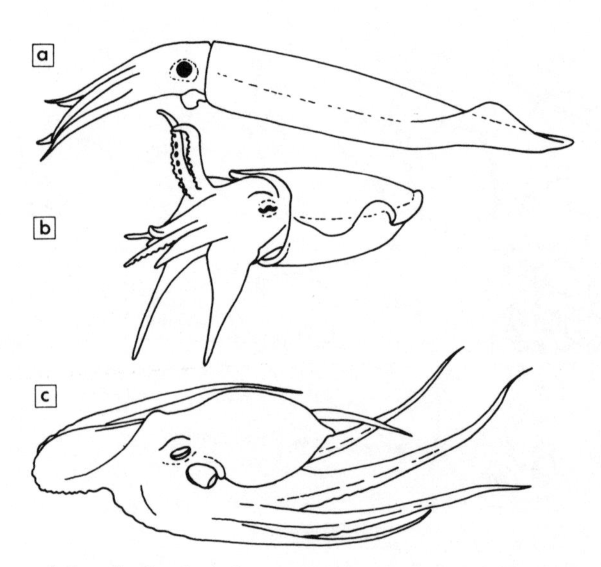

Fig. 2. Three cephalopods that have been widely used for research in the laboratory: A, squid; B, cuttlefish; and C, octopus.

mantle cavity of the female. In squid and cuttlefish there is, in addition to the eight arms, a pair of tentacles, longer than the arms, that are shot out at great speed to capture prey.

The mouth, lying at the center of the suckered arms, is furnished with a parrotlike beak and a radula for breaking down food (Altman and Nixon, 1970). Only small pieces of food can be ingested, as the narrow esophagus passes through the center of the brain. The secretions of the salivary glands serve to immobilize the prey and to begin digestion. Octopuses feed on a variety of organisms, including other mollusks, marine arthropods, and fish. At one time it was thought that octopuses extracted shelled mollusks by force, but it is now known that the radula is used to bore through the shell so that the saliva can enter to weaken or kill the prey (Arnold and Arnold, 1969; Pilson and Taylor, 1961). The food passes from the esophagus into a crop, where it may be temporarily stored. Subsequently, the food is mixed with the secretions of two digestive glands, the "pancreas" and the "liver," as it passes along the digestive tract. Excretory products may be initially formed in the "liver," where high concentrations of ammonia, urea, and uric acid are found. The products are then extracted from the blood by the pericardial glands, renal organs, and gills. The gills are paired and lie on the outside of the visceral mass within the mantle cavity. Respiratory currents are provided by movements of the muscular mantle that alternately draw water into the cavity through a pair of valves and expel it through the funnel. The oxygen-carrying pigment, hemocyanin is dissolved in the blood. Although its carrying capacity is low (5 vol. percent), oxygen utilization can be as high as 90%.

On the basis of the respiratory system the recent cephalopods have developed jet-propelled locomotion, with the muscular funnel being used to direct the water jet. Octopuses generally jet backward when escaping. Movement in this direction allows them to adopt their most streamlined form with the arms held together and trailing behind the body. Octopuses also "walk" on the tips of their arms and swim slowly backward by opening and closing the arms and associated interbrachial web like an umbrella. When attacking their prey octopuses jet forward with the leading arms curved backward. Attacks are visually controlled by one eye only and may occur at speeds in excess of 1 msec (Maldonado, 1964). Jets of water are also used by octopuses to clean their suckers, to aerate their eggs, and to dissuade approaching "enemies." Ink may be mixed with the jet when an octopus is threatened with attack. The ink particles may either form a dark cloud to mask the retreating octopus or, if mixed with mucus, hang together to distract the aggressor with an octopuslike shape while the real octopus, now changed from a dark to a light color, makes its escape.

Water jets are used by cuttlefish both to bury themselves in sand and to flush their prey from the sand. During hunting, cuttlefish use undulating movements of the lateral fin for gentle, low-speed locomotion. Octopuses have no fins; squids use their fins for floating and slow swimming (Mangold-Wirz, 1969). Unlike the attack of octopuses, the cuttlefish attack is binocularly controlled (Messenger, 1968).

In the faster-moving squids and cuttlefish the mantle muscles, whose contractions produce the water jet for locomotion, are innervated through a giant fiber system. Paired first-order giant fibers arise in the giant fiber lobe of the brain, and the axons run to the palliovisceral center. Here the first-order axons fuse in the midline and then synapse with several second-order giant neurons, which send axons to the funnel and the paired stellate ganglia. Third-order giant fibers run from the stellate ganglia to innervate the mantle muscle. Such a system ensures lateral synchronization and the simultaneous contraction of many muscle fibers, resulting in the short reaction time required for rapid escape.

The success of the cephalopods as marine organisms owes much to their solutions to the problem of buoyancy. They have developed three types of buoyancy mechanisms based on fats, tissue fluids, and gas spaces (Denton and Gilpin-Brown, 1973; Denton, 1974). Some oceanic squid have extraordinarily large livers, which contain much low-density oil, presumably produced by a special method of fat metabolism. Specific gravity can also be reduced by the replacement of heavy ions in the body fluids with lighter ones. Some pelagic squid have large volumes of coelomic fluid in which nearly all of the cations have been replaced by ammonium ions, accumulated as ammonium chloride. Gas-filled spaces in the shells of such cephalopods as the cuttlefish and pearly *Nautilus* illustrate the third mechanism. These animals control their buoyancy by exchanging liquid between the blood and the gas spaces via the siphuncle. The bottom-living *O. vulgaris*, on the other hand, would gain little from being neutrally buoyant, and it has no obvious buoyancy mechanism.

In cephalopod mollusks the sexes are separate. The male octopus produces elaborate packets of sperm, spermatophores, that are passed to the female by a modified arm, the hectocotylus. Copulation usually involves some form of courtship. In the octopus the male, who is usually smaller than the female, approaches his mate carefully with his first and second arms turned back to reveal an especially large pair of suckers on the second and third arm of each side (Packard, 1961). While he is still at a distance, the hectocotylus stretches out and the tip appears to "caress" the female before entering her mantle cavity to deposit the spermatophores. Octopuses lay their eggs in clusters like long bunches of grapes hung side by side. The female remains to guard and care for her eggs, aerating them with jets of water from her funnel and cleaning them with the tips of her arms. During this period, which may last six weeks, she refuses food and loses weight, and usually dies soon after the eggs are hatched. Female *O. vulgaris* probably live for 15 to 24 months (Nixon, 1969). The lifespan of the male cannot be accurately assessed but it is thought to be somewhat longer.

Whereas young cuttlefish emerge from the egg as miniature adults, *O. vulgaris* hatch as planktonic larvae. The young octopuses settle on the bottom some 33–40 days after hatching (Itami *et al.*, 1963). Growth may be rapid with an increase in body weight from 50 g to 2000 g taking about 11 months (Nixon, 1969), but the rate of growth is markedly influenced by temperature (Mangold and Boletzky, 1973).

D. The Major Sense Organs: The Statocysts, Suckers, and Eyes

The cephalopods, like the vertebrates, have developed their paired statocysts for the detection of angular rotation in three planes at right angles to each other. This function is performed by the hair cells of the crista statica, while those of the macula statica serve as gravity receptors. Other hair cells, found on the wall of the statocyst sac, may detect pressure changes or vibrations (Young, 1960*a*; Budelmann *et al.*, 1973). There is evidence that octopuses cannot respond to sounds of 100–2000 cycles/sec, while low frequencies that do elicit responses are responded to equally by octopuses both with and without statocysts (Hubbard, 1960; Dijkgraaf, 1963). Recording from the middle cristal nerve, however, Maturana and Sperling (1963) found the statocyst to be very sensitive to low-frequency vibration (tapping the bench produced bursts of firing) but it gave no response to airborne sounds.

Each of the eight arms of *O. vulgaris* bear some 300 suckers arranged in a double row. Receptors in the suckers provide much chemical and mechanical information. It has been estimated that the arms bear as many as 2.4×10^8 primary receptors, of which there are three main morphological types (Graziadei, 1971). Eighty to ninety percent of the primary receptors are tapered ciliated cells, while the remainder are round cells and irregular multipolar cells. Graziadei (1964, 1965) has suggested that the round and irregular cells may be mechanical receptors, while the more numerous ciliated cells are probably chemoreceptors. The latter are concentrated in the rim and infundibulum of the suckers. *O. vulgaris* is capable of detecting some chemicals at concentrations of 100–1000 times less than the threshold for man (Wells, 1963). In a series of experiments that will be considered in more detail later (section VIA), Wells (1965*b*) has demonstrated the apparent inability of octopuses to make use of proprioceptive information for discrimination learning.

The eyes of modern cephalopods are in some ways strikingly similar to those of vertebrates. The structures are not homologous, however, for whereas the vertebrate eye arises as an outgrowth of the neural tube, the cephalopod eye is formed by an invagination of the ectoderm. Because the cephalopod lens is rigid, with a fixed focal length, accommodation cannot be achieved by alterations in its shape. Instead, contraction of the ciliary muscle moves the lens closer to the retina. There is a closed, fluid-filled chamber behind the lens, but the fluid-filled space in front of the lens is continuous with the orbit and is also open to the exterior through a narrow duct. The iris surrounds a horizontal, slit-shaped pupil, which is usually maintained in a constant orientation with respect to gravity. Control is exercised by a reflex link with the statocysts. The retina is not inverted and the many optic nerves leave the retina from behind, so there is no blindspot. The retina of the octopus contains a regular arrangement of some 25 million rhabdomes, rodlike receptors each composed of four rhabdomeres. The rhabdomes, which are separated by slender pigment cells, are thinnest and most densely packed within a horizontal strip situated just above the equator of the retina. Light adaption is achieved by rapid adjustment of pupil size, change in the length of the receptors, and movement of the screening pigment (Young, 1963*a*).

Sutherland (1963*a*) measured visual acuity by giving transfer tests with stripes of different widths to octopuses of 250–500 g trained to discriminate vertical from horizontal stripes. He estimated that the minimum discriminable stripe subtended an angle of 17′. Packard (1969) calculated that this is equivalent to a spacing of four rhabdomes on the retina, a figure which agrees with his own estimate of three rhabdomes obtained with smaller (less than 22 g) octopuses tested by means of their optokinetic response to rotating striped backgrounds. The weight of behavioral evidence suggests that the octopus does not possess color vision (Messenger *et al.*, 1973), yet remarkably accurate matching of the immediate environment is achieved through nervous control of the pigment-filled chromatophores (Packard and Sanders, 1969, 1971) and passively by reflection from the iridophores and leukophores that lie in the skin (Messenger, 1974).

E. The Nervous System

The cephalopod nervous system probably arose from an arrangement of cords and not by the concentration of separate ganglia (Young, 1971). An idea of the original arrangement may be obtained from the system found in *Nautilus*, where anterior and posterior subesophageal cords unite at the sides of the esophagus and join a supraesophageal cord. During the course of evolution these cords have become divided into distinct lobes that have a central neurophile surrounded by a layer of cell bodies.

The octopus brain is somewhat unusual in that it continues to increase in size and in cell numbers throughout life (Packard and Albergoni, 1970). It is divided into some 64 lobes protected by a cranium composed of tissue resembling vertebrate cartilage. The three primitive cords of the *Nautilus* condition can still be recognized in the octopus brain as the supraesophageal lobes, the anterior subesophageal lobes, and the palliovisceral lobe (Fig. 3). The supra- and subesophageal lobes are connected by numerous nerve bundles running round the sides of the esophagus. Large optic lobes

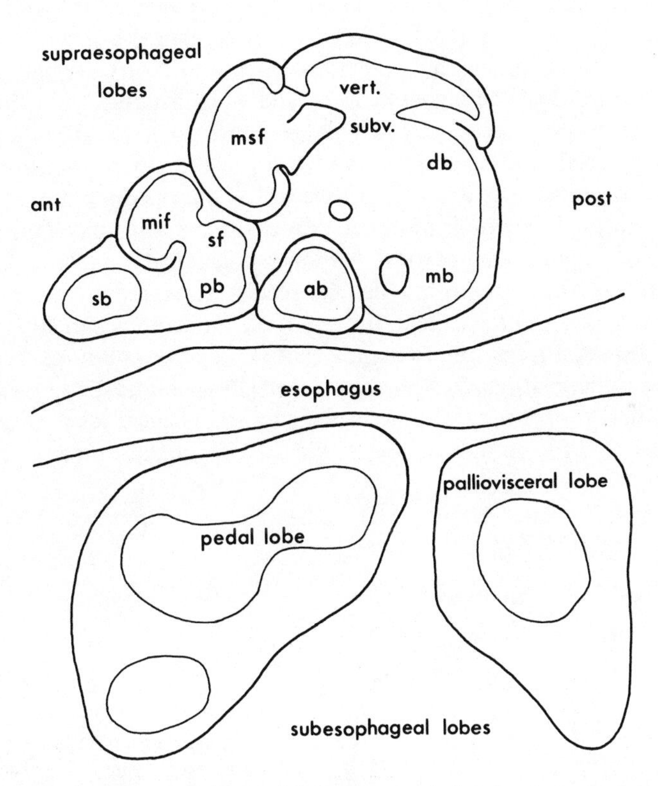

Fig. 3. A sagittal section of the brain of *Octopus vulgaris* showing the major lobes and features visible in this section. Note the three subdivisions that represent the primitive plan of three cords; the supraesophageal lobes, the pedal lobe, and palliovisceral lobe. *ab*, anterior basal; *ant*, anterior; *db*, dorsal basal; *mb*, median basal; *mif*, median inferior frontal; *msf*, median superior frontal; *pb*, posterior buccal; *post*, posterior; *sb*, superior buccal; *sf*, subfrontal; *subv*, subvertical; *vert*, vertical.

lie on each side of the brain between the cranium and the eyes (Fig. 4). These optic lobes have developed from the supraesophageal cord, and they remain attached to the supraesophageal lobes by the optic stalks. Eight brachial nerve cords, one for each arm, arise from the anterior subesophageal mass (Fig. 5). Situated at the center of each arm and running its entire length is the axial nerve cord, which consists of a series of ganglia and a pair of nerve bundles. In addition, each arm contains five other smaller nerve centers, four intramuscular nerve cords, and the ganglia associated with each sucker (Graziadei, 1971). It has been estimated (Young, 1963*b*) that the arms contain nearly 350 million neurons, while the paired optic lobes contain 129 million and the remainder of the brain only 40 million.

The nervous system of the octopus has been classified into various functional centers (Boycott and Young, 1950; Young, 1971). The lower motor centers, e.g., the ganglia of the arms, contain the large final motor neurons that control muscles and other effectors. Normally they operate under the control of the higher centers, and when isolated from the rest of the nervous system, they act only in response to local stimulation. An isolated arm will pass pieces of food from sucker to sucker toward the mouth. The pedal lobes are an example of the intermediate motor centers, which act through the lower motor centers in the arms to control a particular activity, e.g., maintaining posture. Although controlled mainly by the higher motor centers, the intermediate centers receive some input from receptors. When the higher motor centers are removed, an octopus can still move its arms to protect a point on the mantle that is touched, and it can regain the normal upright position after being upturned (Boycott and Young, 1950).

The higher motor centers lie in the supraesophageal lobes (Fig. 3) and are responsible for complex actions. Electrical stimulation has shown that

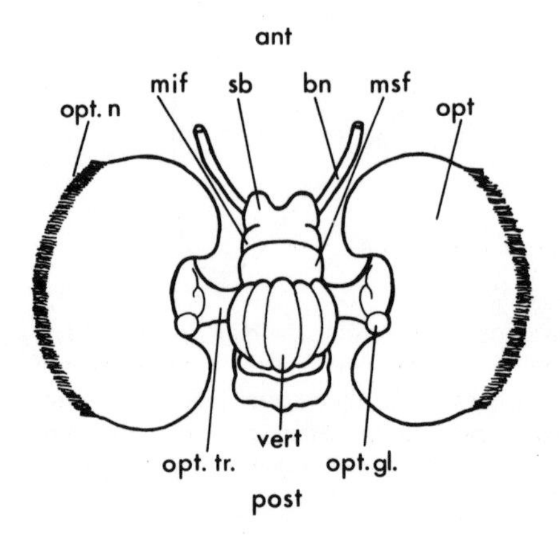

Fig. 4. A diagram of the brain of *Octopus vulgaris* as seen from above, showing the large optic lobes. (After Young, 1971.) *ant*, anterior; *bn*, brachial nerve cord; *mif*, median inferior frontal; *msf*, median superior frontal; *opt*, optic lobe; *opt. gl.*, optic gland; *opt.n.*, optic nerves; *opt.tr.*, optic tract; *post*, posterior; *sb*, superior buccal; *vert*, vertical.

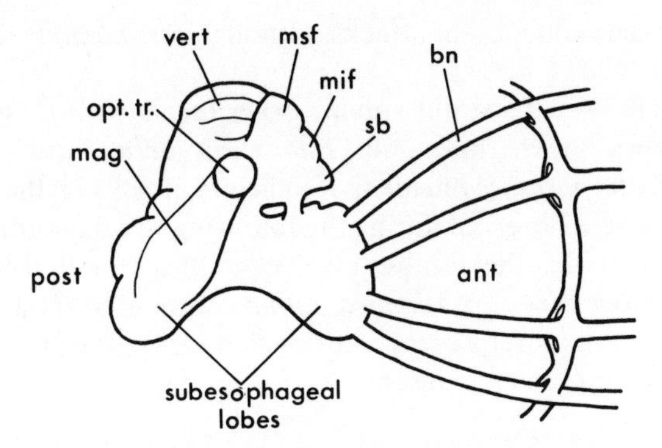

Fig. 5. A diagram of the brain of *Octopus vulgaris* as seen from the right side, showing the origins of the brachial nerve cords that serve the arms. (After Young, 1971.) *ant*, anterior, *bn*, brachial nerve cord; *mag*, magnocellular; *mif*, median inferior frontal; *msf*, median superior frontal; *sb*, superior buccal; *opt. tr.*, optic tract; *post*, posterior; *vert*, vertical.

the anterior basal lobe controls walking movements, the median basal lobe swimming, and the lateral basal lobe the color changes seen in the skin (Boycott, 1961; Young, 1971). Two other functional groups have developed from these higher motor centers. Belonging to one type are the receptor analyzers such as the optic lobes for vision and the inferior frontal–posterior buccal lobes for touch. The other centers are those that appear to be predominantly associated with learning and memory, of which the median superior frontal and vertical lobes are prominent examples. Much of the experimental evidence for the functions of these lobes will be discussed later in this chapter.

IV. EARLY STUDIES OF LEARNING

A. Avoidance of Sea Anemones

One of the earliest investigations of learning in cephalopods used a problem that was commendably close to one the subject might be expected to meet in the wild. After starving *Eledone moschata* (the lesser octopus) for 15 days von Uexkull (1905) presented them with hermit crabs carrying sea anemones. The *Eledone* made several attempts to capture the crabs but retreated on being stung by the anemones. Von Uexkull did not test the *Eledone* with other prey, but he reported that they never attacked the hermit crabs again. Polimanti (1910) repeated this experiment and reported a far less marked reduction in attacks, although one *Eledone* did not attack for a period of 24 hr.

Boycott (1954) carried out similar experiments with *O. vulgaris*. Two octopuses shown the hermit crab, *Eupaguras bernhardus*, carrying the anemone *Callactis parasitica* made five and eight attacks on the first day. No further attacks were observed during the following five days although the octopuses readily took other crabs, bivalve mollusks, and fish. When *Eupagurus prideauxii* carrying *Adamsia palliata* were used, however, the outcome was different: after several curtailed attacks, the octopuses finally ignored the anemone and ate the crab.

B. Habituation

Goldsmith (1917*a,b,c*) reported a series of experiments with octopuses, including two studies of habituation. One octopus attacked a source of agitation in the water 15 times without reward before it stopped. One or two hours later it attacked only twice before stopping. In a second experiment she investigated habituation with an inedible object. On the first presentation the octopus passed the object to its mouth and held it for a long time. After another six trials the object was seized but quickly rejected. Buytendijk (1933) obtained habituation of the withdrawal response to a white card 10 × 10 cm. He passed the card over the octopus at 1- to 2-sec intervals. Initially the octopus reacted violently by withdrawing its head, blinking, and moving away from the stimulus. After 15 trials the withdrawal response had habituated and the octopus was attending to the card with its head bobbing gently and inclined toward the stimulus. Five minutes later six trials were required to achieve the change from withdrawal to attention, whereas, after a further 5 min, withdrawal occurred on the first trial only.

C. Conditioning

If gently disturbed, *Eledone* produces two dark eye spots on the upper surface of its mantle. Mikhailoff (1920) used this response to demonstrate conditioning. He showed that the unconditioned stimulus (gentle poking with a glass rod) reliably produced the eye spots while the conditioned stimuli (red, green, and white lights) did not. Successful conditioning of the eye-spot response to red and green lights was achieved but attempts with white lights failed. The failure with white light is perhaps not surprising because the experiments were not conducted in the dark. Using beating with a stick as the unconditioned stimulus, Kuhn (1930, 1950) reported the conditioning of an avoidance response in octopuses to different wavelengths and intensities of light. Both of these experiments were insufficiently controlled for their results to be considered as evidence of color vision (see Messenger *et al.*, 1973).

D. Discrimination Learning

Goldsmith (1917*a,b,c*) investigated discrimination learning by presenting octopuses with mussels held in colored forceps. After an octopus had eaten a mussel taken from green forceps, she presented it with two pairs of forceps, one green and the other yellow. The octopus seized the green forceps that had been associated with the mussel. After four such trials she demonstrated retention for periods of 3–4 hr. In other experiments similar results were obtained when mussels associated with colored discs were presented to the octopus. Goldsmith concluded that octopuses could discriminate green, yellow, blue, red, and black; however, these experiments, like those of Mikhailoff (1920) and Kuhn (1930, 1950), were insufficiently controlled to be accepted as evidence for color vision (see Messenger *et al.*, 1973).

Ten Cate and Ten Cate-Kazeewa (1938) fed an octopus six to ten times a day in association with a square of lead 10×10 cm. After a week the octopus attacked the square when it was placed in the water alone. At this stage they continued feeding in association with the square but also presented a triangle without food. After 54 unrewarded attacks on the triangle, the octopus stopped attacking it although it continued to attack the square.

Boycott (1954) tried to teach octopuses to discriminate between a crab shown alone and a crab shown with a white 5-cm square. Five trials, two positive and three negative, were given each day at approximately 3-hr intervals. On the positive trials the octopuses were allowed to capture and eat a crab shown alone. On the negative trials, which lasted 4 min, the crab was presented with the white square and pulled away as the octopus attacked only to be replaced at the other end of the tank. The seven octopuses trained in this way made from 18 to 35 (median 21) attacks on the crab and square during the 4-min period of the first negative trial. A criterion of 5 attacks or less over 5 successive trials was reached in 2–13 trials (median 9).

E. Detour Experiments

Early attempts to train octopuses in simple detours were largely unsuccessful. Bierens de Haan (1949) reported that an octopus presented with a detour around a piece of wire netting for a crab reward persisted in trying to reach the crab through the netting and never detoured. He also mentions Buytendijk's repetition of this experiment, in which one of several octopuses finally went around the wire netting to capture the crab, but no further details are given. Boycott (1954) failed to train five octopuses to run through a chicane-type detour for a crab reward. He also reports failure to train three octopuses to make an immediate detour around a glass partition to capture a crab. However, one of the three was given four trials per day for 11 days. As a result the time spent pushing at the glass before detouring was reduced from a mean of 26 sec on day 1 to 9 sec on day 11.

Schiller (1949) used a more extensive detour apparatus (Fig. 6). The octopus was faced with two chicken-wire windows separated by the entrance to an opaque corridor. A crab placed behind one of the windows was visible to the octopus and could be obtained by a detour through the corridor followed by a turn to the correct side. During the course of training, with random positioning of the crab, the time spent trying to reach through the window decreased and the number of successfully completed detours increased. A typical performance over the first 15 trials was 1 outcome ambiguous, 7 wrong, and 6 correct, while the following 10 trials were all correct. It should be noted that the direct attacks at the window never disappeared but that the time spent indulging in this behavior became progressively shorter (cf. Boycott, 1954).

Schiller found that the octopuses were capable of correct choices after delays of up to 1 min. He reported that this success resulted from maintaining the majority of the suckers in contact with the wall that separated the octopus from its prey. On ten trials when he disrupted this

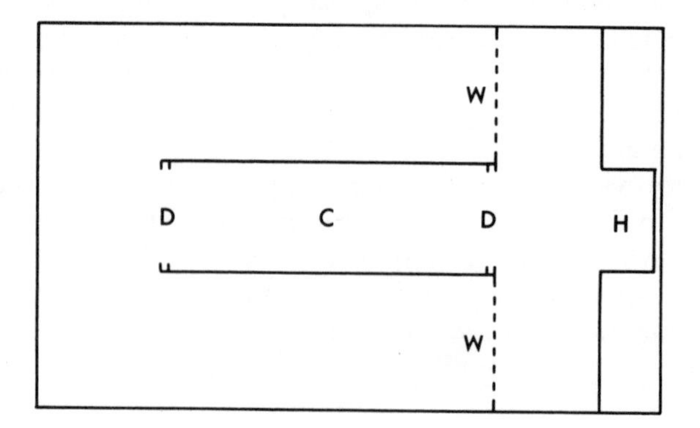

Fig. 6. A plan of the detour apparatus used by Schiller (1949) and Wells (1964*c*, 1967, 1970). From its home (*H*) the octopus sees a crab presented behind one of the windows (*W*). To capture the crab the octopus must detour through the opaque corridor (*C*), in which it may be detained by the closing of the guillotine doors (*D*).

positioning by forcing the octopus to crawl through a small hole, only four out of eight completed responses were correct: a chance level of performance.

F. Problem Boxes

Several investigators have tried to train octopuses to take food from inside glass or metal containers. Pieron (1911, 1914) presented an octopus with a crab in an aspirator that had the smaller of its two openings closed with a rubber bung. The octopus frequently removed the bung but did not immediately reach for the crab. Pieron reported that the futile attacks against the glass decreased and using the arms to feel inside the aspirator began after two weeks of training and became firmly established after 24 days' training. This response was extinguished in a week and reestablished in three days. The report lacks detail and it is not clear whether the octopus learned to feel inside the aspirator, as the observed reduction in attacks against the glass may have resulted in an increase in chance entries by the arms. Certainly, Boycott (1954) was unable to train an octopus to remove a rubber bung in order to seize a crab.

In a similar experiment Boycott (1954) showed a crab on the rim of a metal pot, allowing it to fall inside the pot as soon as the octopus moved forward to attack. On early trials the octopuses initially searched behind the pot, whereas later they attacked the pot directly, but during 80 to 100 trials with ten octopuses, no systematic change to searching inside the pot with the arms was observed.

Schiller (1948) presented six *O.vulgaris* with seven different problems involving a crab in a container. In five of these problems the octopuses easily obtained the crabs but in two no consistent improvement occurred. A floating container was occasionally pulled down and the crab taken, but these successes did not result in an increase in the frequency of this response. Again, when the octopuses were presented with a choice of containers, the baited one was not touched more frequently than the empty one.

G. The Early Learning Experiments of Boycott and Young

Extensive investigation of learning in octopuses began with the work of Boycott and Young. Many of their publications concerned the effect of vertical lobe ablation on learning and memory, an aspect of the work that will be considered later in this chapter. The spate of publications that followed the appearance of "The comparative study of learning" (Boycott and Young, 1950) has been summarized by Wells (1965*a*).

In the sea, octopuses are solitary animals that make a home in a suitable shell, pot, or holes in rock, from which they emerge to hunt for food. Boycott and Young based their training procedure on this behavior. In the laboratory the octopuses were isolated in opaque aquaria provided with a home at one end. The training differed from previous procedures in the use of both positive and negative reinforcement to establish a discriminated response. The positive and negative stimuli were presented separately on successive trials with a food reward for attacks on the positive stimulus and a small (5–12 V a.c.) shock for attacks on the negative stimulus.

When octopuses have adapted to laboratory conditions they readily attack both living crabs and dead sardines when these are presented alone (Boycott and Young, 1955*a*). Discriminated responding was rapidly established when the octopuses were allowed to take and eat one of the stimuli, while a shock was given for attacks on the other (Table I).

Boycott and Young (1950) used a crab shown alone as the positive and a crab plus a 5-cm white square as the negative stimulus. Six trials, three positive and three negative in a quasi-random sequence, were given each day at 2-hr intervals. Each trial lasted 2 min. Positive trials terminated when the octopus captured the crab. If the octopus attacked the crab and white square on negative trials it received an 8-V a.c. shock through electrodes attached to the rear of the square. On being shocked the octopus returned to its home, but the negative stimulus was left in the tank for the full 2 min of the trial, so the octopus could, and occasionally did, make more than one attack during a single negative trial. Attacks on the positive stimulus remain fairly constant over the first 20 positive trials. Four octo-

Table I. Total Attacks Made Over 15 Pairs of Trials with Crabs and Sardines

(Data from Boycott and Young, 1955*a*)

Discrimination	Octopus	Positive stimulus	Negative stimulus
	1	15	4
Crab +ve/fish −ve	2	14	4
	3	12	2
	4	9	4
Fish +ve/crab −ve	5	14	5
	6	14	2

Table II. The Performance of Six Octopuses Trained with Crab Alone as the Positive Stimulus and Crab plus Square as the Negative

(Data from Boycott and Young, 1955*b*)

	Number of octopuses making an attack			Median attack latency (sec)	
Day	Trials	+ve	−ve	+ve	−ve
1	1	6	6	3.5	8.5
	2	4	4	16.0	100.0
	3	4	0	47.5	—
2	1	5	1	29.5	—
	2	6	0	4.5	—
	3	6	0	13.5	—
3	1	5	0	18.0	—
	2	6	0	5.0	—
	3	6	1	4.5	—

puses made 20 attacks on the first block of ten positive trials and 21 during the second block of ten trials. Attacks on the negative stimulus, however, showed a marked decrease, with 34 occurring during the first block of ten negative trials but only 6 during the second. The results of a similar experiment (Boycott and Young, 1955*b*) are summarized in Table II.

Boycott and Young (1956) demonstrated discriminated responses to different shapes by presenting the two discriminanda separately on successive trials, both accompanied by a crab. The octopus was allowed to eat the crab from the positive stimulus complex, but attacks on the negative stimulus complex resulted in the delivery of an 8–12-V a.c. shock, causing the octopus to retreat without the crab. Differential responding is reported for squares of different sizes, squares and diamonds, squares and rectangles, squares and crosses, and horizontal and vertical rectangles, but not squares and circles.

V. TYPES OF LEARNING

A. Habituation

Wells and Wells (1956) extended the work of Goldsmith (1917*a,b,c*) on the habituation of a response mediated by the octopus's tactile system. If a small inedible object such as a Perspex cylinder or sphere, with or without grooves cut into its surface, is placed within reach of an octopus (or allowed to touch the arm of a blind octopus), it will be passed toward the mouth. The object will be examined for a period of about 2 min before it is rejected (Wells and Wells, 1956, 1957*a*). On repeated presentation, at an interval of 2 min after rejection, the response rapidly changes. Within five to nine trials the octopus ceases to pass the object toward its mouth, while over the next ten trials the time spent examining the object with an arm tip before rejection decreases progressively to a base level of about 4 sec. When the taking response has been reliably habituated to one object, others that differ sufficiently in texture will be taken as soon as they are presented for the first time.

The reports of Buytendijk (1933) describing habituation of the withdrawal response to large novel stimuli have been discussed in section IV, B.

B. Conditioning

There have been very few investigations of conditioning in cephalopods that have employed standard classical or operant techniques. Two studies of classical conditioning, Mikhailoff (1920) in *Eledone* and Kuhn (1930, 1950) in *Octopus*, have been described in section IV, C.

Dews (1959) reported the conditioning of lever movement in *O. vulgaris*, but this procedure appeared to be restricted to continuous reinforcement because the octopuses held the lever, making a discrete response difficult to obtain. Crancher, *et al.* (1972) attempted to reproduce similar conditioned lever-pulling in *O. cyaneus* but did not succeed. They did, however, obtain successful conditioning of another free operant: the octopus was required to extend an arm up a tube that dipped vertically into the water of the tank. Of six octopuses, one produced the operant spontaneously at a base rate of about twice per 30-min session and one required shaping for three 1-hr sessions, while the remainder learned the response within one 30-min shaping session. Considerable variability of response rate was found between octopuses with continuous and fixed-ratio (FR2) schedules of reinforcement. Continuous reinforcement produced a steplike pattern of responding, but a consistently high rate of responding was obtained with a variable-ratio schedule (VR2). It should be noted, however, that the amount of data collected was small.

C. Associative Learning

Octopuses will not readily attack stimulus objects until they have been trained to do so. Although feeding before presentation of the stimulus increases the frequency of attack, the effect is short-lasting; more long-lasting changes are obtained only when food reinforcement is given after the attack has occurred (Young, 1960*b*). The change in the frequency and latency of attack are not merely a function of acclimatization to laboratory conditions. Octopuses that have been trained to attack a black vertical rectangle with latency of a few seconds also attack on most tests with crabs and other shapes but always after much longer delays, while octopuses trained to attack crabs will not necessarily attack plastic shapes (Young, 1956). Even the latency of attack on crabs shows decreases that cannot be attributed entirely to acclimatization (Maldonado, 1963*a*).

If octopuses are given a small electric shock (4–12 V a.c.) for attacks on a particular stimulus, the frequency of attack rapidly decreases and remains low for some time, whereas the effect of shock given before presentation of the stimulus is short-lasting (Young, 1956, 1960*b*). Young (1970) trained 11 octopuses not to attack a crab by giving an 8-V a.c. shock for each attack with trials spaced at 5-min intervals. The attacks fell from 100% in the first three trials to remain below 30% for trials 7 to 12. On the following day, 7 hr after a further 12 training trials, only 3 of the 11 octopuses attacked the crab, whereas 9 attacked a vertical rectangle, suggesting that the learning was specific to crabs.

Sanders and Barlow (unpublished data) compared the effect of no shock and shocks of 4 V and 10 V on the crab-attack latency of octopuses, with trials given at intervals of approximately 3 sec. Representative cumulative latency curves showed a marked difference between the no-shock and shock conditions, with a more rapid increase in attack latency occurring with the higher voltage (Fig. 7). It is apparent that the reduced attack rate cannot be attributed to exhaustion and/or habituation. It is also unlikely that the shocks themselves caused the decrease in attack rate because five 10-V shocks given alone at intervals of 1 sec before the first presentation of the crab did not affect the response rate (Fig. 7).

We have seen (section V, A) that octopuses will readily pass small objects toward the mouth for examination and that, in the absence of reward, this response rapidly habituates. In their early studies of tactile learning Wells and Wells (1956, 1957*a*) demonstrated that the response of taking a particular inedible object could be reliably established when each take was reinforced with a small piece of fish. Learning *not* to make this response occurs particularly rapidly (Wells, 1959*a*) if a 6-V a.c. shock is given each time an octopus takes the object. With trials at 5-min intervals seven octopuses made from 2 to 4 takes before reaching a criterion of 5 sucessive rejections. Similar results were obtained when the trials were spaced at 20 min (Wells, 1959*b*).

Maldonado (1968) investigated active-avoidance learning in octopuses using a shuttlebox. The animals were trained to move over a hurdle from a

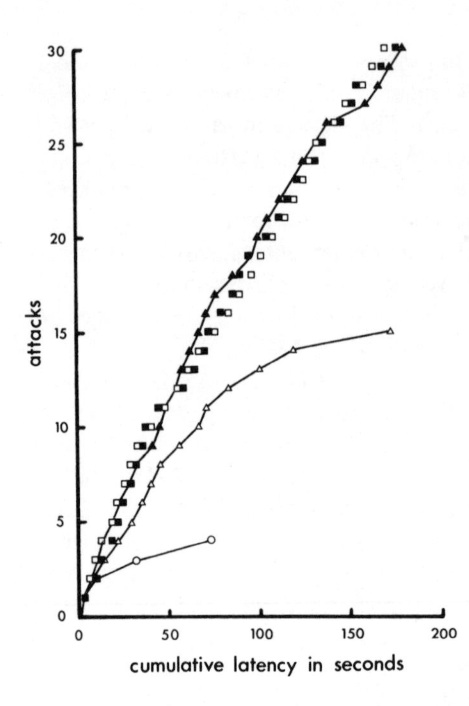

Fig. 7. Cumulative latencies of attack on a crab obtained with five different conditions: no shock (*open squares*), probe without shock (*filled squares*), 4-V shock for each attack (*open triangles*), 10-V shock for each attack (*open circles*), and a shock control (*filled triangles*) given five 10-V shocks, at intervals of 1/sec, immediately before the first presentation of the crab. Each curve represents the performance of one octopus that was typical of the condition. The octopuses were never allowed to capture the crab, which was removed at each attack and immediately replaced 50 cm from the octopus toward the other end of the tank. (Sanders and Barlow, unpublished data.)

darkened box to a brightly illuminated box in order to avoid a pulsed 10-V shock. An avoidance, or anticipatory run, was scored each time an octopus left the dark box within the first 10 sec of darkness before the shock began. Incorrect crossings were scored each time an octopus returned from the light to the dark box. One trial terminated and the next began when the octopus had spent 25 sec in the light box. The animals were given 20 trials in a single session. There was no difference between the numbers of anticipatory runs in the first and second blocks of ten trials, but the number of incorrect crossings decreased significantly from a mean of 3.4 to 1.4. A second experiment (Maldonado, 1969) confirmed these findings and in addition showed that although there was no reduction in the latency of anticipatory runs (avoidances), the "error time" (the time spent being shocked in the dark box) was significantly reduced. It appears that octopuses cannot readily learn an active avoidance to the onset of darkness, but they do learn to passively avoid the dark box in which they have been shocked.

Octopus maya has been trained to traverse a 1.5 m runway for a food reward (Walker, *et al.*, 1970). Three animals given five trials per day at intervals of 1 hr reduced their median time per trial from near 200 sec to an asymptote of 50–60 sec in about 14 days. These authors were also successful in training *0. maya* on a spatial discrimination in a simple T-maze.

Motivation was provided by removal of the seawater, and the correct response, a left or right turn from the stem of the maze, was reinforced by return to seawater. Five animals, each trained against a slight preference, significantly reduced their errors over a period of 27 days when given three trials per day at intervals of 1 hr (Fig. 8). In future this maze could be modified for use with either visual or tactile discrimination tasks.

Sanders and Young (1940) described the waning of the attack response in cuttlefish (*Sepia*) when prawns were presented behind glass. The attack may be divided into three sequential stages: (1) attention, when the cuttlefish turns to face and fixate its prey; (2) positioning, when the cuttlefish moves towards its prey, stopping at the correct attacking distance from it; and (3) seizure or strike, when the pair of long tentacles are shot out at great speed to capture the prey. Such a discrete response as the strike is particularly suited to studies of learning. When prawns are presented behind glass, cuttlefish make repeated attacks, striking their tentacles against the glass. The response rate decreases gradually until a criterion of no strikes within 3 min is reached in a period of 20–30 min. Some savings were made in retraining after retention intervals of up to 18 hr.

Recently, Messenger (1973*a*) has extensively studied this phenomenon. He reported that the three responses constituting the attack—attending, positioning, and striking—wane at different rates during a continuous 20-min period of prawn presentation (Fig. 9). The strike rate falls to less than a quarter of its initial level during the period from 5 to 10 min. The frequency of positioning does not fall below half its initial value until the period from 10 to 15 min, while attending has fallen to only 75% of its initial level by the 15–20-min period. Waning of the response is specific to prawns because all of the cuttlefish would subsequently attack crabs.

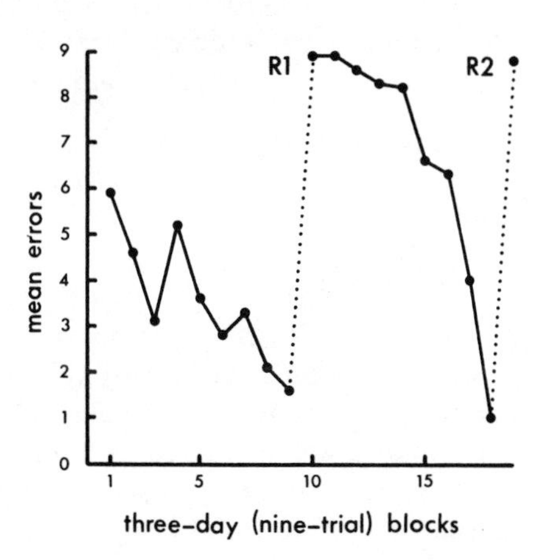

Fig. 8. The acquisition and reversal of a spatial discrimination in a simple T-maze as measured by the mean errors recorded by 5 *O. maya*. The chance level of performance is 4.5 as each point represents nine trials given at the rate of three per day. *R1* and *R2* indicate the start of the first and second reversals. (Data from Walker *et al.*, 1970.)

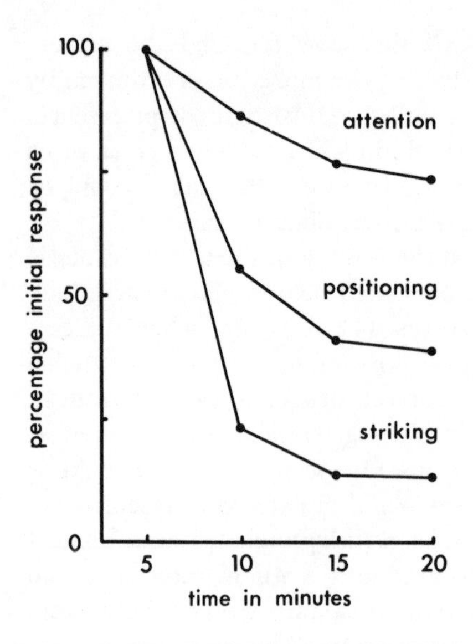

Fig. 9. The waning of the three stages of attack in cuttlefish ($N = 30$) during a 20-min period of continuous presentation of a prawn behind glass. The performance over each 5-min block is expressed as a percentage of that obtained over the initial 5 min. (Data from Messenger, 1973*a*.)

Although the waning of the cuttlefish's attacks on the prawn behind glass has a superficial resemblance to the phenomenon of habituation it is better classified as associative learning. Sanders and Young (1940) noted that the striking of the tentacles against the glass may be painful, and Messenger (1973*a*) has now clearly demonstrated that the rate of waning can be influenced by changes in the level, or type, of reinforcement. He compared the performance of the following groups of animals:

Group 1, 10 cuttlefish that were given a 10-V a.c. shock for each strike at a prawn behind glass.
Group 2, 30 cuttlefish allowed to strike without receiving shock.
Group 3, 10 cuttlefish from which the tentacles had been surgically removed to eliminate pain input from tentacles striking the glass.

With the tentacles removed, the cuttlefish still showed attention and positioning like normal animals and pseudostrikes could still be observed because their bodies made a small, but obvious, forward jerk as they attempted to strike.

Cuttlefish from the three groups were each exposed to the prawn behind glass for 20 min. The rates of striking were different for the three groups, with the shocked cuttlefish always attacking less than the normal cuttlefish, which, in turn, attacked less than those without tentacles. When performance was measured in terms of a percentage of the initial level of attacking, the frequency of striking was found to decrease fastest in group 1 and slowest in group 3. Messenger (1973*a*) successfully prevented waning of the attack in another group of nine cuttlefish by giving a dead prawn as a reward after each block of five strikes at the prawn behind glass. Finally, if normal cuttlefish are prevented by a transparent screen from getting within striking distance of the prawn, they never attempt to strike. Such exposure, for periods of 0, 6, 12, or 18 min, subsequently had no effect on the initial frequency of striking at the prawn behind glass nor on the rate of learning.

D. Visual Discrimination

The earlier investigation of visual discrimination learning employed large food rewards (a whole crab or 6 g of sardine) for attacks on the positive stimulus, making relatively long intertrial intervals of 1–2 hr essential (Boycott and Young, 1956; Sutherland, 1957*a,b,* 1958*a,b,;* 1959*a;* 1960*a*). Subsequent studies used smaller fish rewards and shorter intertrial intervals of predominantly 5 min (Sutherland, 1959*b;* 1960*b,c;* 1961). The shapes have generally been presented on transparent rods and moved up and down gently by hand through a distance of about 5 cm three times per sec. Stationary shapes are not readily attacked by octopuses and presentation by machine produces results similar to those obtained by hand presentation (Sutherland, 1957*a*).

Training by successive presentations of the discriminanda with food reward for attacks on the positive stimulus and small shocks for attacks on the negative has been predominantly used with octopuses. This method works well for easy discriminations, but with difficult discriminations there is a decrease in the number of attacks made, which may fall as low as 10% of the trials, so that much of the time is wasted and interpretation of the performance level may be difficult. With simultaneous training, level of performance is independent of level of attack. Early attempts to train octopuses by means of simultaneous presentation were not successful (Boycott and Young, 1956; Sutherland and Muntz, 1959) but Maldonado (unpublished data) showed that acquisition would occur, providing that shocks were given for attacks on the negative stimulus as well as food reward for attacks on the positive. Subsequently, a few studies have employed this method of training. Muntz *et al.* (1962) used it to investigate the effects of vertical lobe removal (see section VII, E). Rhodes (1963) successfully trained octopuses on a simultaneous size discrimination with a 5-cm-diameter disc as the positive and a 6-cm-diameter disc as the negative stimulus. An asymptote of 90% correct responses was reached after 70 trials given at the rate of 3 per day at intervals of 3–4 hr. Sutherland *et al.* (1963*a*) compared the two methods of training and found that with easy discriminations the same asymptote was reached but that with difficult discriminations performance was better when the stimuli were presented simultaneously.

Octopuses often show marked tendencies to attack one member of a pair of shapes rather than the other. A knowledge of such preferences is important in the design and interpretation of visual discrimination experiments. It is well established that octopuses prefer vertical to horizontal rectangles when both are moved vertically (Young, 1958*a*, 1965*a*). Sutherland and Muntz (1959) showed that this preference is reversed when the rectangles are moved horizontally. They also report a preference for shapes that move in the direction of their points (cf. Sutherland, 1959*b*). Small shapes are preferred to large and black is preferred to white when the stimuli are shown against a cream-colored background, but the latter preference disappears, or is reversed, with a gray background (Young, 1958*a*, 1968).

The relationship between brightness and orientation was investigated by Messenger and Sanders (1972). They recorded the numbers of attacks made by 36 octopuses on stimuli presented against a cream-colored background in a series of unrewarded tests. The marked preference for vertical over horizontal when the rectangles are moved vertically was confirmed, but an interaction between orientation and brightness was also apparent (Table III). The orientation preference, which is most marked with black rectangles, almost disappears when the rectangles are white. Black is attacked more than white if the rectangles are vertical, but the reverse is true of horizontal rectangles. When orientation was eliminated by the use of circles, black was attacked more than white, with 55 to 37 attacks respec-

Table III. Total Number of Unrewarded Attacks Made by 36 Untrained Octopuses on Six Rectangles Each Presented Twice Against a Cream-Colored Background

(Data from Messenger and Sanders, 1972)

	Brightness of the rectangles			
Orientation of the rectangles	Black	Light gray	White	Probability of an attack
Vertical	51	43	32	0.58
Horizontal	17	28	28	0.34
Probability of an attack	0.47	0.49	0.42	0.46

Table IV. A Comparison of Level of Learning and Level of Attack over Trials 49–64 on a Black/White Brightness Discrimination Using Either Vertical or Horizontal Rectangles as the Discriminanda

(Data from Messenger and Sanders, 1972)

	Vertical		Horizontal	
Positive stimulus	Black	White	Black	White
Percent correct attacks	82	91	81	80
Mean attacks on the positive	4.9	5.5	2.2	2.7

tively. It would appear that a marked preference exists for rectangles with particular orientations. The probability of an attack on a particular shape may be further influenced by its brightness insofar as brightness affects the discriminability of the shape from the background. In discrimination training, preference was shown to have a marked effect on the level of attack but little effect on the amount of learning (Table IV).

Boycott and Young (1957) trained octopuses with pairs of discriminanda that differed in orientation, brightness, size, or shape and obtained performances of better than 70% correct responses over 120 trials (60 positive and 60 negative) for all except a square/diamond discrimination (Table V). Although very few animals were trained on each discrimination, the range of performance on the same task was small.

Sutherland (1957*a*) found that although octopuses could readily learn to discriminate between a pair of rectangles oriented at 90° to each other when they were presented with one horizontal and one vertical (horizontal and vertical rectangles), they could not discriminate these shapes when they were presented after turning through 45° (opposite obliques).

The discrimination of vertical and horizontal rectangles is dependent on their orientation with respect to the retina, not gravity. The slit-shaped pupil of the octopus is normally held horizontally, but after bilateral statocyst removal the reflex control of eye orientation is lost (Boycott, 1960; Young, 1960*a*). Orientation of the retina with respect to gravity now depends on the position of the animal, but the octopus continues to behave as if the retina were correctly oriented (Wells, 1960). Octopuses that had previously been trained to attack vertical and not to attack horizontal rectangles performed correctly after statocyst removal only when the pupil

Table V. Performance of *O. vulgaris* over 120 Trials on Various Visual Discrimination Problems

(Data from Boycott and Young, 1957)

Discrimination	N	Percent correct responses
	3	88 (range 86–90)
	2	88 (range 86–96)
	1	77
	3	72 (range 71–73)
	1	61

happened to be held horizontally. On trials when the eye was at right angles to the normal orientation, the octopuses tended to attack the negative vertical rectangle and to avoid the positive horizontal rectangle.

Opposite obliques are a special example of left-right mirror-images, and, as such, they may generate confusion when viewed binocularly (Mello, 1965*a,b;* Noble, 1966). When octopuses were trained to discriminate opposite obliques monocularly, however, no improvement on a chance level of performance was obtained, although vertical and horizontal rectangles were readily discriminated (Messenger and Sanders, 1971). Possible neuroanatomical correlates of these behavioral findings have been described by Young (1960*c;* 1962*a*). Cells in the plexiform layer of the optic lobe have oval rather than circular dendritic fields with particular orientations. The majority have the long axis of the field horizontal but some are arranged vertically. These cells may serve as detectors of excitation in the vertical and horizontal planes.

On the basis of his finding that octopuses can discriminate between a horizontal and a vertical rectangle but not between opposite obliques, Sutherland (1957*a*) suggested that these animals classify shapes according to their vertical and/or horizontal extents (see Table VI, problems 1–4). To test these alternatives Sutherland (1957*b*) trained octopuses on problems 5 and 6 of Table VI. Problem 5 is soluble in terms of horizontal extents and problem 6 in terms of vertical extents. On finding that the octopuses performed equally on the two problems, he proposed a theory of shape discrimination based on the utilization of both horizontal and vertical extents. These results were the start of a series of experiments designed to discover the mechanism by which octopuses make visual discriminations of shape and orientation. A test of this hypothetical model is summarized in Table VII. The model predicts that octopuses should be able to solve problems *a* and *b* with equal facility, whereas problems *c* and *d* should be insoluble. In fact, performance on problem *a* was significantly better than on problem *b* ($P < 0.01$). It is also reported that significantly more correct than incorrect responses were made on problem *c*, but the test used, a chi-squared test on the total number of responses, is inappropriate. Transfer tests indicated that shapes M and Λ are very similar for octopuses, but the different levels of performance obtained on problems *a* and *b* demonstrate that these shapes are discriminable. Sutherland (1959*b*) suggested that octopuses may

Table VI. Performance of *O. vulgaris* over 60 Trials on the Discrimination of Pairs of Shapes Differing in Orientation Only

(Data from Sutherland, 1957*b*)

Problem	Discriminanda	Horizontal extent	Vertical extent	N	Percent correct responses
1	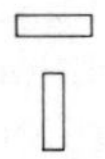	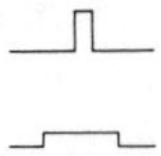	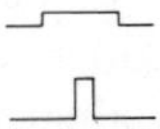	6	81[a]
2	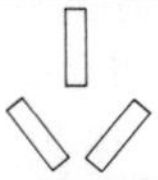	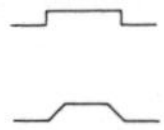	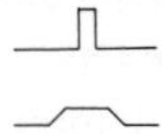	8	71[a]
3	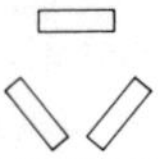	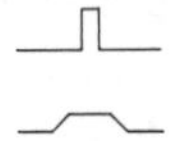	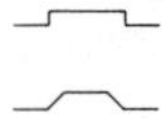	8	65.5[a]
4		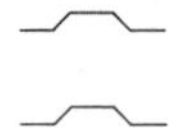		6	50
5	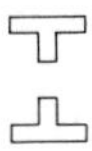	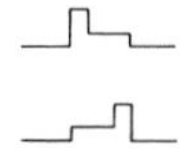	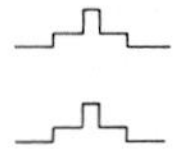	6	59[a]
6		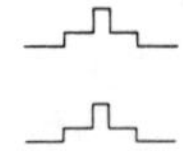	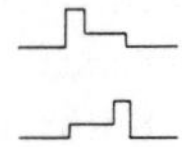	7	56[a]

[a] Better than chance level of performance $P < 0.05$.

Table VII. Visual Discrimination Performance of *O. vulgaris* over 1280 Trials Related to the Horizontal and Vertical Extents of the Discriminanda

(Data from Sutherland, 1959*b*)

Problem	Discriminanda	Horizontal extent	Vertical extent	*N*	Percent correct responses
a.				8	69
b.	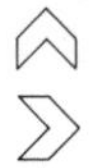	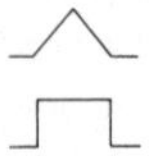		8	56
c.		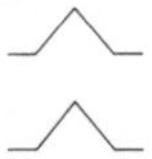	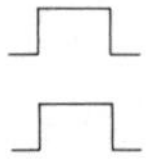	8	53
d.	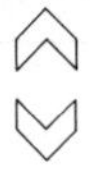	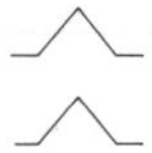	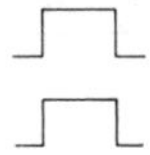	8	50

distinguish M and Λ either by analyzing them in relation to a fixation point or by using additional mechanisms of shape discrimination. Further discussion of Sutherland's original theory may be found in Deutsch (1960*a,b*) and Sutherland (1960*d,e*). Failure of the theory to predict these and other experimental results, in particular the finding (Sutherland *et* al., 1963*b*) that octopuses could discriminate between the two stimuli illustrated in Figure 10a, led to the proposal of a new theory of shape recognition (Sutherland, 1963*b,* 1969) based on elongated visual receptive fields of the type described by Hubel and Wiesel (1962) in cats, and essentially similar to those suggested for octopuses by Young (1960*c*).

A total of six distinct theoretical models of shape discrimination in the octopus have been published. In addition, Young (1960*c,* 1962*a*) has described cells in the optic lobes of octopuses that have oval dendritic fields with predominantly horizontal and vertical orientations and has suggested that they may form part of a system for computing the horizontal and vertical extents that form the basis of some of these models. Deutsch (1955) suggested a mechanism based on computing the distance between the contours of the shape. Sutherland (1957*b*) proposed that shapes were analyzed by conversion to their horizontal and vertical projections. Deutsch (1960*a*) outlined a second theory according to which octopuses calculate the reciprocal of the vertical distances between contours at all points on the horizontal axis. Sutherland (1960*d*) added another dimension to his original theory, suggesting that octopuses also compute the length of contour of a shape divided by its area. Dodwell (1961) proposed that the direction of the contours of a shape were analyzed with reference to the horizontal or vertical axis. Finally, Sutherland (1963*b*) suggested that shapes were recognized by elongated receptive fields.

To test these theories Muntz (1970) devised the pair of training shapes *A1* and *A2* shown in Fig. 10b. Ten octopuses were given 16 training trials per day in two blocks of 8 (4 positive and 4 negative) for seven days. Performance over the last five days averaged 63% correct and was significantly different from chance. None of the above theories could clearly predict this result except the first proposal of Deutsch (1955). According to this theory the shapes are discriminable because the horizontal distances between the contours are not all the same. However, this hypothesis would also predict that the transfer shape *B4* (Fig. 10B) would be indistinguishable from *A2* because the distances between the contours are the same in these two shapes, yet in transfer tests *B4* was treated like *A1*. This finding does not necessarily invalidate the theories because it is probable

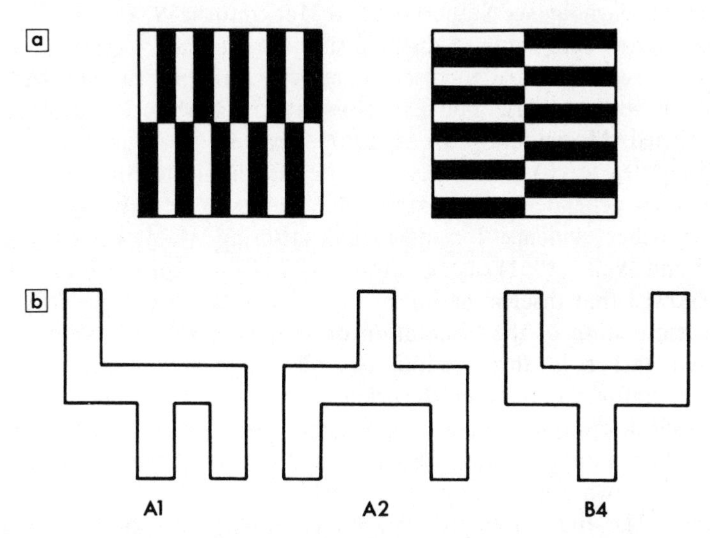

Fig. 10. a, Two visual training stimuli used by Sutherland *et al.* (1963*b*) that were readily discriminated by octopuses. Each 10 × 10 cm square was composed of reduplicated 5 × 1 cm rectangles cut from black and white Perspex. b, The two visual training stimuli (*A1* and *A2*) and one of eight transfer stimuli (*B4*) devised by Muntz (1970) to test six theories of shape discrimination in octopuses. The stimuli were 10 cm long horizontally and cut from white Perspex.

the octopus uses more than one mechanism for shape recognition. Sutherland (1968) has distinguished between global and component descriptions of shape. Except for Sutherland's last theory, which involves component descriptions, all of the above proposals depend on global descriptions. The results obtained by Muntz suggest that component rather than global descriptions are used without the loss of information relating the component parts to the whole shape.

Eninger (1952) showed that rats trained with two relevant cues reached criterion faster than they did with only one of the cues present. This phenomenon, known as "additivity of cues," has been demonstrated in octopuses by Messenger and Sanders (1972). Octopuses trained with both orientation and brightness relevant performed consistently better than others trained with either one alone, although the difference in performance diminished after the first 32 trials. Transfer tests indicated that the problem was solved predominantly in terms of one cue or the other with 22 octopuses utilizing brightness, 6 using orientation, and 1 showing no difference in its transfer scores. Despite the tendency to solve the problem in terms of one cue, most octopuses learned something about both cues. These findings are in agreement with those obtained by Sutherland and Holgate (1966) with rats and are readily explained by a two-process model of discrimination learning (see Sutherland and Mackintosh, 1971).

As well as being able to distinguish different shapes, there is evidence that octopuses can learn to discriminate the direction of polarization of light; both vertical/horizontal and oblique (45°/135°) orientations can be distinguished (Moody and Parriss, 1961). Transfer from one source of polarized light (a torch) to another (a white disc) indicates that the discrimination is not dependent upon extraocular scattering or reflections (Moody, 1962). Further evidence for intraocular discrimination was provided by Rowell and Wells (1961), using octopuses after bilateral statocyst removal. They showed that discrimination of the orientation of polarized light, like the discrimination of the orientation of rectangles (Wells, 1960) that was discussed earlier in this section, depends upon the maintenance of a constant relationship between the orientation of the retina and the stimuli.

Visual discrimination learning has also been demonstrated in *Sepia* by Sanders and Young (1940), who trained three cuttlefish to discriminate between a prawn and a prawn plus a white circle. Four trials were given each day. The first three were negative trials, in which the prawn was presented behind a glass plate bearing the white circle. The last trial each day was a positive, "feeding" trial, in which the cuttlefish were allowed to capture and eat a prawn shown alone. The criterion of learning was six consecutive negative trials, during the first 5 min of which no attacks were made, accompanied by two positive trials, in which the prawn was taken.

The three cuttlefish reached this criterion after 38, 48, and 55 negative trials.

E. Discrimination by Touch

Blind octopuses can be trained to discriminate by touch between living lamellibranch mollusks and wax-filled shells of the same species (Wells and Wells, 1956). Species whose shells differed in texture could also be distinguished. Wells and Wells's most widely used tactile-training procedure employed small Perspex cylinders with and without 1-mm-wide grooves cut into the surface. Subsequently the cylinders were replaced by similar spheres in order to present a uniform radius of curvature. Training involved successive presentations of the positive and negative stimuli to the suckers at the distal end of one arm. The most widely used criterion of a "take" was the passing of the cylinder under the interbrachial web toward the mouth. Small pieces of fish were given to reinforce takes of the positive stimulus. Takes of the negative stimulus were immediately followed by a 6–9-V a.c. shock delivered to the arms or mantle through two electrodes mounted on a probe.

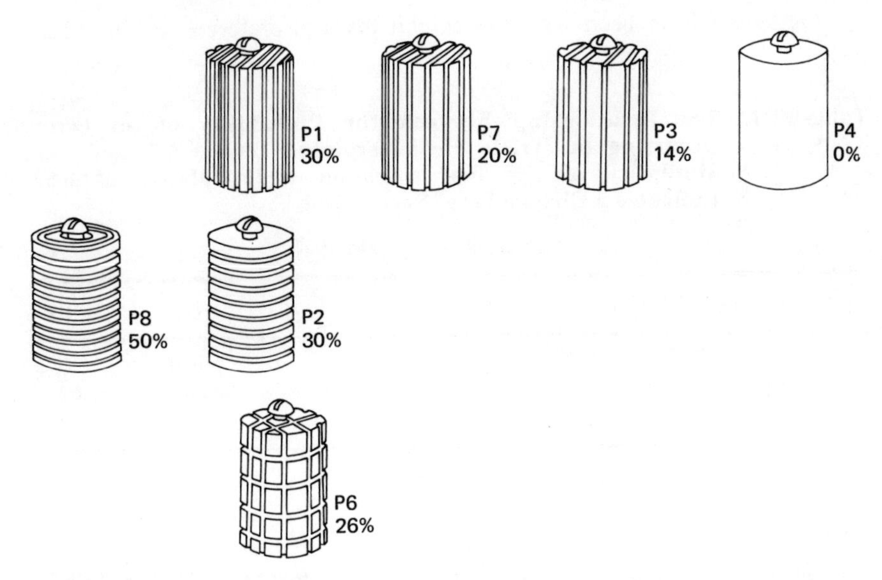

Fig. 11. The four types of Perspex cylinder used by Wells and Wells for tactile discrimination experiments with octopuses. One is smooth, while the others have grooves 1 mm wide and 1 mm deep cut into the surface. A graded series of textures was produced by variation of the distance between the grooves. The percentages beside each stimulus indicate the proportion of the surface area cut by grooves. (After Wells and Wells, 1957*b*.)

Nine octopuses, given four positive and four negative trials per day at minimum intervals of 1 hr, rapidly learned to discriminate a smooth cylinder from one with grooves cut 3 mm apart. A criterion of 75% correct responses over the eight trials that constituted one day's training was reached in a mean of 4.3 days. In a total of 468 pairs of positive and negative trials, the octopuses made 437 takes of the positive cylinder and 213 takes of the negative.

By using a series of cylinders (Fig. 11) with different percentages of their surface area cut by grooves, Wells and Wells (1957*b*) were able to demonstrate that the difficulty of the task was inversely proportional to the difference between the discriminanda in terms of the percentage of the surface area cut by the grooves (Table VIII). This relationship remained true when the discriminanda differed markedly in the pattern of grooves Fig. 11. The cylinder *P1*, which had only longitudinal grooves, was not discriminated from *P2*, which bore circumferential grooves, nor from the *P6* combination of longitudinal and circumferential grooves (Table VIII and Fig. 11). It would appear that the octopus is not capable of integrating proprioceptive information about the relative positions of its suckers with the information about texture received from them in order to differentiate patterns of grooves.

Octopuses have been shown to exhibit textural preferences, but these

Table VIII. The Relationship Between the Probability of an Error Occurring on Trials 57–96 (Training Days 8–12) and the Similarity of the Discriminanda—A Probability of 0.50 Indicates a Chance Level Responding

(Data from Wells and Wells, 1957*a*)

Discrimination	P4/P8	P4/P1	P4/P7	P1/P3	P1/P6	P1/P2
Types of grooves	none/ circum	none/ longit	none/ longit	longit/ longit	longit/ longit and circum	longit/ circum
Difference in surface area cut by grooves	50%	30%	20%	16%	4%	0%
Number of octopuses	12	12	6	5	6	6
Median probability of an error	0.08	0.14	0.31	0.34	0.41	0.49

are not generally so marked as visual preferences (see section V, D) in unoperated animals (Wells and Young, 1968*a*). Over 160 extinction trials with rough and smooth spheres, ten octopuses made a mean of 35 takes of the smooth and 23 of the rough sphere. This preference for smooth was seen in discrimination training in which octopuses trained rough-positive/smooth-negative initially performed at a lower level than those trained smooth-positive/rough-negative, but the asymptote reached was the same for both (cf. Messenger and Sanders, 1972; section V, D).

VI. OTHER LEARNING PHENOMENA

A. Proprioception and Learning

From the failure of octopuses to perform learned visual discriminations of orientation correctly after bilateral statocyst removal (Wells, 1960; Rowell and Wells, 1961) and their inability to discriminate by touch between a cylinder with longitudinal grooves and one with circumferential grooves (Wells and Wells, 1957*a*), it appears that octopuses cannot utilize proprioceptive information for discrimination learning. Further evidence for this view has been provided by experiments on the discrimination of weight and shape. Wells (1961*a*) trained four blind octopuses to discriminate by touch between two smooth cylinders that were identical in all characteristics but weight. One weighed 5 g and the other 45 g in seawater. During training the arms of the octopuses could be seen to compensate for the weight of the cylinders, but over 280 trials there was no improvement on a chance level of performance, although on a rough/smooth discrimination, errors fell to less than 10% after only 40 trials. Proprioceptive information was available for control of the arms but not for discrimination learning.

The discovery that blind octopuses could learn to discriminate a cube from a sphere by touch (Wells, 1964*a*) appeared to contradict this view. However, when the corners of the cube were rounded off the number of errors increased, and a rod, having the same radius of curvature as the corners of the rounded-off cube, was treated like the cube. Further experiments with cylinders of different diameters and composite cylinders composed of bundles of rods (Wells, 1964*b*) demonstrated that the cues for discrimination learning came from the degree of distortion of individual suckers and not from proprioceptive information about the degree of curvature of the arm. Composite cylinders were treated as being of the diameter of their component rods. In transfer tests with rough and smooth cylinders of equal, large diameter, five octopuses that had been trained to accept a smooth cylinder of large diameter and to reject one of small diameter took the smooth 48 times and the rough only 13 times. Clearly, a rough cylinder of large diameter is treated as a smooth cylinder of small diameter. It appears that the tactile system of *Octopus* discriminates forms by the same mechanism that is used for texture: the proportion of receptors that are excited when the suckers touch the stimulus.

B. Reversal Learning

Single reversals have been reported for octopuses trained on both visual (Young, 1956, 1962*b*; Boycott and Young, 1958) and tactile (Wells and Wells, 1957*c*) tasks, and Walker *et al.* (1970) obtained a single reversal of a spatial discrimination with *O. maya* in a simple T-maze (Fig. 8). Young (1962*c*) trained octopuses for eight days with 20 trials per day on three visual discrimination tasks, reversing the discrimination at the start of each day. In terms of errors per trial the performance became progressively worse, but this was the result of a continued decrease in the numbers of attacks. When the proportion of correct attacks to total attacks was used as a measure the performance of animals trained on a black/white discrimination and that of others trained with a square/diamond problem showed an improvement during each day's training and possibly also over successive reversals. This improvement was not seen with animals trained on a horizontal/vertical-rectangle task. As the octopuses were not trained to criterion but for a fixed number of trials, failure to find improvement in the last problem probably reflects the strong preference that octopuses show for vertical over horizontal rectangles when both are moved vertically.

Mackintosh (1962) trained 18 octopuses to a criterion of 80% correct over the 20 trials of one day's training on a vertical/horizontal-rectangle problem. Before each reversal of the problem half of the animals were given 20 and half 60 overtraining trials. From 2 to 9 reversals were completed, but no evidence was found for an improvement in the learning of successive reversals nor for the overtraining reversal effect (ORE), an improvement in reversal learning resulting from overtraining that is typically found with rats (for a review see Mackintosh, 1965*b*). The failure to find an improvement in this experiment cannot be attributed to the strong preference for vertical rectangles, because training to criterion was employed and comparisons were made between alternate reversals so that performance on the same problem was compared.

These two repeated reversal experiments with octopuses used successive training, whereas most of the demonstrations of progressive improvement and ORE in rats used simultaneous training and therefore, of necessity, position was an irrelevant cue. To investigate the effect of irrelevant cues, Mackintosh and Mackintosh (1963) compared the reversal performance of three groups of octopuses trained to a criterion of 90% correct on a black/white discrimination when half of the animals in each group had been given 100 overtraining trials and the other half had been reversed without overtraining. Group I (no irrelevant cues) was given successive training on the simple black/white discrimination. Group II (position irrelevant) was given simultaneous training on the black/white problem. Group III (orientation irrelevant) was given successive training with black and white rectangles that were either horizontal or vertical according to a random sequence. The ORE appeared in the reversal performance of groups II and III, which had been trained with irrelevant cues, but not in group I (Table IX).

The demonstration of an ORE in the presence but not in the absence or irrelevant cues is consistent with Sutherland's (1959*c*) two-process model of discrimination learning, which involves attention to the relevant cue followed by attachment of the correct responses. (For a full account of the role of selective attention in animal discrimination learning see Sutherland and Mackintosh, 1971.) This theory also predicts that a progressive improvement in reversal learning will occur over a series of reversals provided the problem contains irrelevant cues. Mackintosh and Mackintosh (1964*a*) trained ten octopuses on a black/white discrimination by simultaneous presentation. Six animals were trained with oblique rectangles with the orientation of the oblique changing randomly from trial to trial. As discrimination between opposite obliques is apparently impossible for octopuses (Sutherland, 1957*a*; Messenger and Sanders, 1971), position and

Table IX. The Effect of Overtraining on Learning the Reversal of a Brightness Discrimination Learned with and without the Presence of Irrelevant Cues as Measured by Errors to Criterion

(Data from Mackintosh and Mackintosh, 1963)

Irrelevant cues (group)	Errors to criterion on learning the reversal		Significance of difference
	Overtraining	No overtraining	
None (I)	14.25	21.0	$P > 0.1$
Position (II)	15.5	39.5	$P < 0.01$
Orientation (III)	20.5	37.75	$P < 0.02$

not orientation was the effective irrelevant cue. Four other octopuses were trained with black and white rectangles that were changed randomly from horizontal to vertical between trials so that both position and orientation were effective irrelevant cues. From six to ten reversals were completed and significant improvement in performance over reversals 1–6 was found. As predicted, the octopuses trained with two cues irrelevant made more errors over the initial reversals (1–4) but subsequently performed slightly better than those trained with one irrelevant cue. Three animals that reached reversal 9 after training with a single irrelevant cue were then trained with the additional irrelevant cue, but no impairment of their performance was observed. Thus the presence of irrelevant cues hindered octopuses on early reversals but had no adverse effect when introduced on the later reversals. The authors concluded that one of the causes of progressive improvement is that in later reversals animals are able to ignore irrelevant cues. These findings are consistent with Sutherland's theory. Further support for a two-process model based on selective attention is provided by the finding that overtraining, which is known to improve reversal learning, impairs performance on a nonreversal shift from an orientation to a shape discrimination (Mackintosh and Mackintosh, 1964*b*).

If, as has been suggested (Gonzales *et al*., 1967; Bitterman, 1968), another cause of progressive improvement in reversal learning is the increase in proactive interference of one problem on the next, producing increased forgetting, then it should be possible to demonstrate improvement with successive training in the absence of irrelevant cues. Sanders (unpublished data) trained 16 octopuses over a series of four reversals of a tactile discrimination using successive presentations. Half of the octopuses were initially trained rough-positive/smooth-negative and half smooth-positive/rough-negative because octopuses without brain lesions tend to prefer the smooth sphere. Training was continued to a criterion of 80% correct over two successive sessions of 16 trials each. Retention was tested by one retraining session five days later and the problem was reversed on the following day. From the first reversal (*R1*) to the fourth (*R4*) there was a significant drop in the number of responses required to reach criterion, but there was no improvement beyond the score for the original problem (*RO*) on the basis of this measure (Fig. 12). The results of the retention tests show the expected effect of proactive interference with a progressive fall in the percent savings with each reversal, which correlates with the level of performance at the start of each new problem (Fig. 13).

Because each problem is begun at a different level of performance, mean responses to criterion are not a good measure of the rate of learning. The first training session at which an octopus scores 50% correct responses

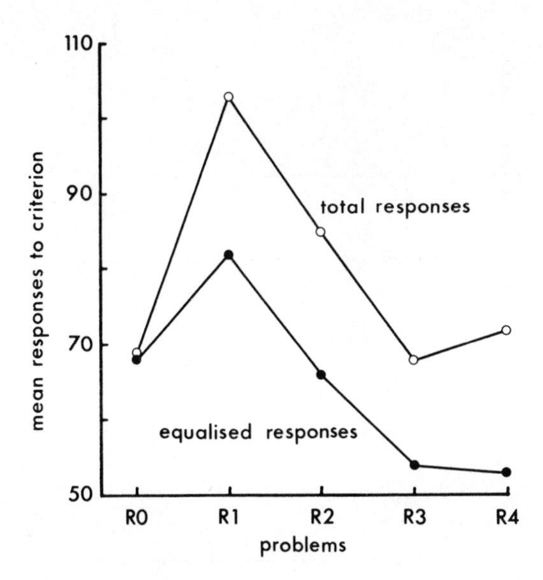

Fig. 12. The progressive improvement shown by 16 octopuses on a series of reversals of a rough/smooth tactile discrimination task. Original problem (*RO*) and four reversals (*R1–R4*). *Open circles* represent the mean total responses to criterion for each problem. *Filled circles* represent the mean "equalized" responses, i.e., the mean responses to criterion as calculated from the beginning of the first training session, at which an octopus scores 50% correct responses or better. The equalized responses, because they measure performance from the same arbitrary starting point, are a more accurate measure of the rate of learning. (Sanders, unpublished data.)

may be taken as an arbitrary starting point that represents equal performance for all animals in all the problems. We can obtain a better measure of the rate of learning by calculating responses to criterion from this point. Using this measure (equalized responses to criterion), we find that the performance on *R4* is better than that on the original problem (Fig. 12). Proactive interference could account for improvement up to the level of the performance on the original problem but not beyond. The most likely contender for improvement beyond this level is learning to ignore irrelevant cues, which could produce an improved rate of learning as opposed to the improved starting level that results from forgetting the previous problem. Although no irrelevant cues were added to the discriminanda, one possible irrelevant cue was present in the training procedure used. Both rough and smooth spheres were suspended by thin nylon line, which the octopuses sometimes touched. Wells (1964*b*) has shown that thin rods are treated as rough stimuli in transfer tests. It appears, therefore, that both stimuli were contaminated by an element of irrelevant roughness that

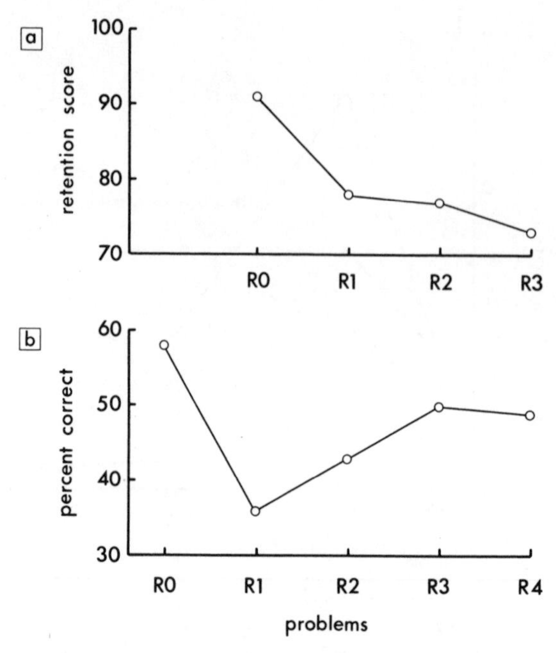

Fig. 13. A, The retention performance of octopuses on a series of reversals of a tactile discrimination compared with B, the level of performance on the first session of the new problem. Retention performance was measured by retraining five days after reaching criterion on the problem, and the reversal began on the following day. The retention scores are offset to facilitate the comparison. Original problem (*RO*) and four reversals (*R1–R4*). (Sanders, unpublished data.)

could account for improvement beyond the original rate of learning if the octopuses came to ignore this feature.

C. Generalization and Transfer

Wells and Young (1970*a*) used tactile training to measure stimulus-generalization gradients in octopuses. The stimuli were a series of fourteen 3-cm-diameter spheres roughened by the presence of 1-mm-wide annular grooves. The series ranged from zero to 13 grooves. Transfer was tested after discrimination training to criterion with a pair of stimuli chosen from this series. After training with the extreme stimuli (zero and 13 rings), identical generalization gradients were found for the two directions of training ($0^+/13^-$ and $13^+/0^-$) with more takes of the intervening stimuli than would be predicted by a linear relationship based on the number of rings. After training with a pair of intermediate stimuli (4 and 7 rings), the generalization curve was found to peak on the side of the positive stimulus that was away from the negative stimulus, a phenomenon that is characteristic of vertebrate generalization curves. The results also indicated that octopuses are capable of quite fine tactile discriminations. During extinction tests with a range of stimuli after training with $7^{+}/13^{-}$, the 7-ring sphere was taken 56 times while the 8-ring sphere was taken only 22 times.

Muntz (1962) trained octopuses monocularly to attack either a vertical or a horizontal white rectangle and tested both eyes with the training shape and with the shape not used for training (the transfer shape). He found the stimulus-generalization gradient obtained with the untrained eye was less steep than that of the trained eye. Both eyes, however, showed an equal increase in the total number of attacks, indicating that there had been a genuine loss of discriminability between the stimuli after interocular transfer (Table X). Interocular transfer of discrimination training was found to be perfect when performance with the trained eye was above 80% correct, but if performance with the trained eye was below this level, transfer was incomplete (Muntz 1961*a*). Muntz obtained a lower level of performance with the trained eye by (1) using distortions of the training shapes as transfer stimuli; (2) using more difficult discriminations; (3) using a lower point on the learning curve; and (4) using a lower point on the extinction curve. All four methods resulted in incomplete interocular transfer.

By presenting a white rectangle, either horizontally or vertically, in a specific part of the visual field for 5 sec before shocking the octopus in its home, Muntz (1963) was able to restrict the avoidance training to one part of the retina. He tested the effectiveness of the training by presenting a crab with the original training shape or with a transfer shape. The stimulus-generalization gradient was steeper when the shapes were presented to the center of the visual field than when the periphery of the retina was used, indicating that the shapes were discriminated more accurately at the center of the retina. In addition, intraretinal transfer was found to be more complete than interocular transfer.

Table X. Total Numbers of Attacks Made on the Training and Transfer Shapes Using the Trained and the Untrained Eyes

(Data from Muntz, 1962)

	Attacks on the training shape	Attacks on the transfer shape	Attacks on both shapes
Trained eye	307	215	522
Untrained eye	268	249	517

Sutherland *et al.* (1963*a*) have demonstrated the phenomenon of transfer along a continuum (Lawrence, 1952) using a square (*S*) with a series of five parallelograms (*P1–P5*), ranging from *P1*, which was least like a square, to *P5*, which was most like a square (Fig. 14). Sixteen octopuses were given ten trials per day with simultaneous presentation of the shapes. Eight received 17 days' training on the difficult *S/P5* discrimination, while the other eight began with 10 days' training on the easy *S/P1* problem. Performance on days 8–10 was 84% correct for the *S/P1* problem and 57% correct for the *S/P5* problem. On days 11–13 the second group were switched from training on *S/P1* to training on *S/P2, S/P3,* and *S/P4,* with 1 day on each problem before completing 4 days' training on the difficult *S/P5* problem. The performance of the second group on *S/P5* over days 14–17 was better than that of the first group, which had been trained on *S/P5* from day 1: the performances were 73% and 54% correct, respectively. In a second experiment (Sutherland *et al.*, 1965) octopuses were trained on a visual discrimination with either size or shape relevant and the other cue irrelevant. The two groups of animals were then trained on a problem that was soluble in terms of either size or shape. Subsequent transfer tests showed that the octopuses initially trained to use size and to ignore shape learned less about shape in the second discrimination than those initially trained to use shape and to ignore size. These findings were interpreted as supporting Sutherland's (1959*c*) theory that animals must first learn to select the relevant analyzer for distinguishing between the stimuli before learning to attach the correct responses to them.

Rhodes (1963) investigated transposition using a simultaneous size discrimination. Six octopuses given five transfer tests each chose the original positive training stimulus 22 times and the relational transfer stimulus 8

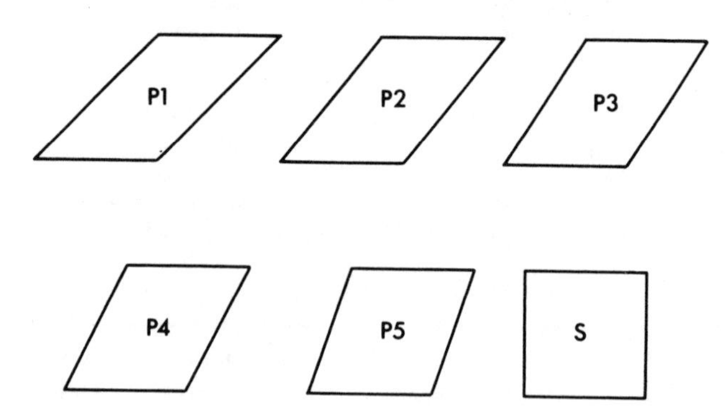

Fig. 14. The series of visual training stimuli—five parallelograms and a square—used by Sutherland *et al.* (1963*a*) to demonstrate transfer along a continuum. Each stimulus was cut from white Perspex and had a surface area of 25 cm².

Table XI. Transfer of Tactile Training in *Octopus* as Measured by Performance on the First Test Trial with an Untrained Arm After Training an Arm on the Other Side of the Body Not to Take a Perspex Cylinder

(Data from Wells, 1959*b*)

Training trial interval (min)	Performance on the first test trial	
	Number of octopuses taking the stimulus	Number of octopuses not taking the stimulus
3	12	5
5	4	2
20	2	10

times. Although this result differs from those obtained with vertebrates, who normally choose the relational stimulus, the performance of the octopuses may have been influenced by a preference for the larger of the two stimuli, which would have produced apparent absolute transposition.

Wells (1959*b*) studied the transfer that results from training one arm of a blind octopus not to take a Perspex cylinder. Typically, the trained arm begins to reject the stimulus after four to five trials. The subsequent performance of an untrained arm on the other side of the body was dependent on the rate of the original training. Tests on the untrained arm were given immediately after the octopuses made six successive rejections with the trained arm. The majority of octopuses trained with trials at intervals of 3 or 5 min took the stimulus when it was first presented to the untrained arm; however, in trials at intervals of 20 min the untrained arm typically rejected the stimulus when first tested (Table XI). Wells interpreted these results as showing that transfer takes several hours to occur and, hence, that there are separate neuronal fields for each arm.

D. Delayed Response and Delayed Reinforcement

Schiller (1949) found that with enforced delays of 1 min an octopus successfully negotiated a detour apparatus on eight out of ten trials, but if the orientation of its body was disrupted during the detour, choices fell to a chance level (see section IV, E). He concluded that the delayed-detour performances in octopuses is mediated by postural cues.

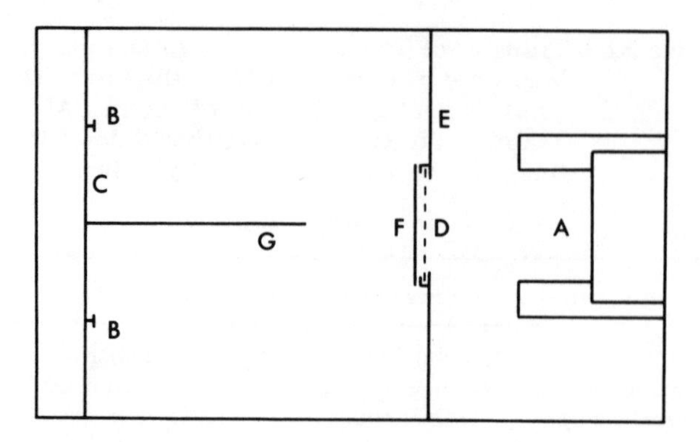

Fig. 15. A plan of the delayed-response apparatus used by Dilly (1963) and Sanders (1970*a*). In its home (*A*) an octopus sees a crab presented beside one of two white rectangles (*B*) attached to a black Perspex screen (*C*). The octopus is prevented from attacking by the transparent guillotine door (*D*) in the black screen (*E*). During the delay period an opaque screen (*F*) is placed against the door and the crab is removed. At the end of the delay the opaque screen and the transparent door are removed together, allowing the octopus to attack. The central division (*G*) prevents the octopus from attacking both rectangles at the same time.

Dilly (1963) used a different apparatus (Fig. 15) to investigate the same problem. He presented the octopus with two identical vertical white rectangles attached to a mat black screen. A longitudinal division placed between the shapes prevented the octopus from touching both rectangles simultaneously. An opaque transverse screen with a transparent guillotine door at its center was placed between the stimuli and the octopus in its home. At each trial a crab was shown beside one rectangle for 30 sec. The octopus was prevented from completing its attack by the transparent door. For delay periods of 10, 20, and 30 sec, during which the crab was removed, an opaque shutter was placed in front of the transparent door. At the end of the delay both door and shutter were removed together and the octopus allowed to attack one of the rectangles. Correct attacks, to the side where the crab had been shown, were rewarded with 0.5 g of fish, while incorrect attacks were punished with 8-V a.c. shocks. At the shorter delays almost all of the trials resulted in attacks and the majority of these were correct. At 10-sec delay there were 21 correct out of 24 attacks in 28 trials. With the 20-sec delay there were 20 correct attacks out of the 26 made in 28 trials. With the 30-sec delay there was some evidence that the octopuses had to learn both to attack a rectangle after the delay and to attack the correct one. The number of trials on which attacks occurred increased from 39% in the first session, through 55% and 65%, to 67% in the fourth session. Similarly, the percentage of correct attacks increased from 57%, through 73% and 92%, to 100% in the fourth session. Dilly checked the possibility that the octopuses were adopting some orientation of the body to mediate the correct response by moving the opaque shutter rapidly toward the transparent door at the start of the delay period. This procedure caused the octopus to retreat from the door to its home but did not disrupt the performance. One octopus made eight correct attacks and twice failed to attack in ten trials with this disruption procedure. Sanders (1970*a*) repeated this experiment without the negative reinforcement for incorrect choices and extended the findings to much longer delay periods. His results for a 30-sec delay agreed well with those of Dilly, but the failures to attack were reduced, perhaps because no shocks were given. The data for longer delays suggest that octopuses are capable of responding correctly after delays of up to 2 min but that after a delay of 4 min the performance falls to a chance level (Table XII).

Wells has studied delayed responses in the detour situation devised by Schiller (1949), in which an octopus is required to make a detour along an opaque corridor in order to get a crab that it has seen through a window to either the right or the left of the entrance to the corridor (Fig. 6). Eleven octopuses, each tested for 20 trials, made 197 correct responses with only 10 errors and 13 failures to complete the detour (Wells 1964*c*). Schiller concluded that success at this task depended on the octopuses' maintaining bodily contact with the intervening wall of the corridor (see section IV, E), but Wells observed correct responses on trials when the octopuses did not touch the walls of the corridor. Blinding the animals in one eye, however, led to systematic errors. When the crab was presented to the *left* the octopus attacked initially with its left eye, but on entering the corridor it led along the left wall with its *right* eye. It is not surprising therefore, that unilaterally blinded octopuses consistently made more errors when detouring toward their intact side. In 63 trials toward their blind side four octopuses

Table XII. The Performance of Octopuses on a Delayed-Response Task

(Data from Sanders, 1970*a*)

Delay in sec	Failures to attack	Correct attacks	Incorrect attacks	Percent correct attacks	
30	32	34	7	83	(from Dilly, 1963)
30	10	31	7	82	
60	14	28	6	82	
120	21	19	8	70	
240	10	14	12	54	

made only 4 errors and 9 failures to complete the detour, while in 63 trials to their intact sides there were 16 errors and 28 failures to complete. Wells concluded that the octopuses completed successful detours by keeping close visual contact with the wall that separated the octopus from its prey.

The delay between seeing the crab and turning from the corridor to the food compartment could be increased if the octopus was shut in the corridor. Wells (1967) found that the proportion of errors increased progressively as the length of the delay increased, but with delays as long as 2 min performance was still better than 70% correct. With delays of more than 2 min, however, the performance fell to chance level. Very similar results were obtained by Sanders (1970*a*) with a different delayed-response problem (see Table XII).

Wells and Young (1968*b*) studied the effect of delayed reinforcement on a black/white visual and a rough/smooth tactile discrimination. The amount of learning diminished as the length of the delay increased, as illustrated by the results of a brightness discrimination (Table XIII). Clear evidence of learning was found with delays of up to 30 sec. The improvement in the 60-sec delay group was entirely the result of the performance of one octopus. Whereas four octopuses in this group performed at chance level throughout the 160 trials, the other animal made better than 70% correct responses toward the end of training. In tactile training on a rough/smooth discrimination over 160 trials, performance with delayed reinforcement was dependent on which stimulus was positive (Table XIV). With the preferred smooth sphere as the positive stimulus, there was clear evidence of learning with delays of up to 30 sec, but when smooth was the negative stimulus no learning occurred with delays longer than 10 sec. One

Table XIII. The Performance of Octopuses Trained on a Brightness Discrimination for 160 Trials with Delayed Reinforcement

(Data from Wells and Young, 1968*b*)

Number of animals	Delay in sec	Mean percent correct responses	
		Trials 1–160	Trial 128–160
5	0	74	86
5	10	71	83
5	15	64	72
9	30	57	60
5	60	54	58

Table XIV. The Performance of Octopuses Trained for 160 Trials on a Rough/Smooth Tactile Discrimination with Delayed Reinforcement

(Data from Wells and Young, 1968*b*)

Delay in sec	Mean percent correct responses	
	Smooth +ve/rough −ve	Rough +ve/smooth −ve
0	74	65
10	69	60
15	63	53
30	73	56
60	58	55
120	56	55

octopus that was trained against the preference with a 30-sec delay took both spheres 30 times during the first 80 trials, but during the last 80 trials it took the rough sphere 28 times and the smooth only 9 times. At the longer delays some individuals made low-level discriminant scores, but these could have arisen from the use of alternate presentation of the discriminanda coupled with a progressive decline in the probability of a take occurring from the beginning to the end of each training session. These animals showed no consistent improvement from session to session as did the octopuses trained with shorter delays. It appears that octopuses are capable of learning both visual and tactile discrimination tasks with a delay of reinforcement of 30 sec. In view of the performance of rats with delayed reinforcement (Grice, 1948), it is perhaps surprising that octopuses can learn with such a long delay. It is unlikely, however, that the learning was mediated by secondary reinforcement, as the investigators were unable to identify differential cues that could have acted as secondary reinforcers. The ability of octopuses to tolerate relatively long delays is less surprising if we consider that in their natural environment there may be a considerable time-lag between capturing an animal and discovering that it is good to eat.

E. Intertrial Interval and Learning

Young (1961*b*) trained octopuses to attack a vertical white rectangle with trials at intervals of either 5 or 50 min. Over the first eight trials the animals trained at an intertrial interval (ITI) of 5 min made slightly more attacks—7.2 compared with 5.9—but the range of behavior in both groups was great and the difference is not significant. A similar result was obtained with a visual discrimination task: horizontal rectangle positive and vertical rectangle negative (Young, 1960*d*). The probability of an error over the first ten trials was 0.38 with an ITI of 5 min and 0.33 with an ITI of 1 hr. Wells and Wells (1957*c*) report the results of training octopuses on a tactile discrimination with ITIs of 5 min or 1 hr. Although the two sets of data are not strictly comparable because of other slight differences in training, both groups made a mean of seven errors over the first 20 training trials. A somewhat different problem was investigated by Wells and Young (1969*a*), who trained octopuses on a tactile discrimination with an ITI of 5 min and trials given in blocks of 8, 16, or 32, with a minimum interval of 6 hr between blocks. Considered in terms of the change in performance per trial there were no differences between the three groups, but on the basis of improvement per take (only takes were followed by reinforcement) learning was better with trials massed in blocks of 32 (Table XV).

Several studies have shown that the optimum ITI for training rodents on active and passive avoidance problems as well as discriminations for food reward is about 5 min (for references see Barlow and Sanders, 1974). With shorter and longer ITIs acquisition is slower. The rate of acquisition in a passive avoidance task has been investigated in octopuses using ITIs ranging from 30 sec to 10 min (Barlow and Sanders, 1974). The octopuses were trained not to attack a crab by use of a 10-V a.c. shock and a criterion of no attack for 1 min. Clear evidence was obtained that the rate of acquisition is a U-shaped function of ITI, with the best performance occurring at ITIs of 1–2 min (Table XVI).

Table XV. The Performance of Octopuses on a Rough/Smooth Tactile Discrimination Task when Trials Were Given in Blocks of 8, 16, or 32

(Data from Wells and Young, 1969*a*)

Number of trials per block	Percent correct responses	
	Over 160 trials	Over 64 takes
8	76	83
16	73	87
32	77	91

Table XVI. The Number of Trials Required to Reach a Criterion of No Attacks for 1 min when Octopuses Are Trained Not to Attack a Crab with Trials at Different Intervals

(Data from Barlow and Sanders, 1974)

Intertrial interval in minutes	Mean trials to criterion	
	Expt. I	Expt. II
0.05	8.7	9.5
0.5	—	8.6
1	5.7	5.5
2	5.4	6.3
5	7.4	6.3
10	7.7	7.6

F. Retention of Learned Behaviors

Sutherland (1957*a*) trained six octopuses for 7 days with 5 positive and 5 negative trials per day to discriminate a horizontal from a vertical rectangle. On testing retention 27 days later by means of 10 unreinforced trials, he found that performances equaled those of the last day of training, with a total of 25 attacks on the positive and 5 on the negative stimulus. These octopuses were, however, trained on discrimination of opposite obliques during the retention interval.

Retention of a rough/smooth tactile discrimination was tested by Wells and Wells (1958*a*) over periods of five and ten days. Six octopuses (group A) were trained to a criterion of 85% correct responses and four (group B) to a 75% correct criterion and then overtrained by the same number of trials. Retention was tested by means of 20 unreinforced trials. After a five-day retention period, group A made a mean of 3.2 errors compared with 0.3 in their last training trials. Group B, after a ten-day retention period, made a mean of 6.8 errors compared with 2.3 in their last 20 training trials. Most of the errors were made by increased taking of the negative stimulus rather than by failure to take the positive. Discriminant performances could be obtained from octopuses that took both the positive and the negative stimuli on all of the test trials if procedures were used that reduced the level of taking. The authors retested two such octopuses, giving a 6-V a.c. shock for each take no matter which stimulus was involved. During these 20 trials one animal made eight positive and two negative takes while the other made nine and five, respectively.

Wells and Young (1970*b*) used repeated presentations of the stimuli at 5-min intervals to produce discriminant scores in extinction. Octopuses were trained on a rough/smooth discrimination for 1, 4, 16, or 30 trials with each of the two stimuli (i.e., 2, 8, 32, and 60 training trials, respectively). Forty-eight hours later 60 unreinforced trials were given to test retention. Octopuses that had learned and retained the discrimination typically began the extinction series by taking both stimuli, but as takes became less frequent they rejected the ex-negative more than the ex-positive stimulus. Eight training trials, but not two, produced a performance significantly above chance level on the test given two days after training. Thirty-two and sixty training trials produced similar performances, both better than that obtained with 8 trials. Training with smooth rather than rough positive consistently produced a better level of performance (Table XVII).

Sanders (1970*b*) trained 45 octopuses on a rough/smooth tactile discrimination and tested retention by retraining after periods of 10–120 days. A typical negatively accelerating retention curve was obtained, with performance falling 25% in 8 days, 50% in 24 days, 75% in 53 days, and 90% in 96 days. Once again, the retention performance was found to be better when smooth was the positive stimulus. The retention of two visual dis-

Table XVII. The Performance of Octopuses on a Retention Test Given Two Days after Different Amounts of Training on a Rough/Smooth Tactile Discrimination—Retention Measured in Terms of the Transformed Probability of a "Correct" Take (Θ) Calculated from Successive Pairs of Trials on Which Only One Take Occurred. Θ Runs from 0 to 90, with 45 Equal to a Chance Level of Performance

(Data from Wells and Young, 1970*b*)

Training trials	Smooth +ve/rough −ve		Rough +ve/smooth −ve	
	N	Retention score Θ	*N*	Retention score Θ
2	4	53.1	5	42.1
8	12	67.7	12	65.1
32	14	78.0	16	73.3
60	8	80.5	8	72.3

Table XVIII. The Biphasic Retention Performance Observed in Cuttlefish Trained Not to Attack Prawns Presented Behind Glass

(Data from Messenger, 1973*a*)

Retention interval	2 min	4 min	8 min	15 min	22 min	30 min	45 min	60 min	90 min	2 hr	4 hr	8 hr	1 day	2 days
N	15	15	15	15	20	19	20	19	17	14	15	14	14	10
Percent savings	96	98	96	93	77	87	81	97	72	77	72	84	83	70

criminations has also been measured (Sanders 1970*a*). With a size discrimination (5-cm circle positive, 2.5-cm circle negative) there was a 42% performance drop in 55 days. In the second experiment the octopuses were trained with black horizontal positive and vertical negative against a gray background. Thus training was against two strong preferences. Over a retention period of only 25 days performance on this discrimination fell by 54%. It would appear that the strength of the original preference, which has been reversed by the training, markedly affects retention performances.

Messenger (1971, 1973*a*) trained cuttlefish not to attack prawns presented behind glass (see section V, C). Retention was tested in a 5-min retraining session given at various intervals, from 2 min to 2 days, after a 20-min training session. Comparison of the number of strikes during the retraining session with the number made during the first 5 min of training provides a retention score that may be expressed as per cent savings. Surprisingly, a biphasic curve was obtained, with better retention performance at 1 hr than at 22 min (Table XVIII). The increases in retention performance at 8 hr and 1 day were not significant. A further experiment showed that the performance after a 24-hr retention period was dependent on the length of the initial training session, suggesting that retention rather than motivation is the factor that determines the level of the retention performance. Cuttlefish trained and retrained with the tentacles removed, a condition that is thought to reduce the amount of negative reinforcement received in training, showed 40% savings after a 2-hr retention period compared with the 73% obtained with intact control animals. This is another indication that the retention performance is dependent on the experience obtained in training.

A somewhat similar result has been obtained by Sanders and Barlow (1971) in training octopuses not to attack crabs. Measuring retention performance after retention intervals of 0.5–30 hr, they obtained a triphasic

Table XIX. The Triphasic Retention Performance Obtained by Training Octopuses Not to Attack a Crab Using a 4-V a.c. Shock and a Criterion of No Attack for 1 min

(Data from Sanders and Barlow, 1971)

Retention interval (hr)	0.5	1	2	3	6	8	10	12	14	20	30
N	6	6	6	7	6	6	7	6	6	7	6
Percent savings	84	68	83	79	56	48	55	75	85	89	86

curve with significant dips at 1 hr and at 8 hr (Table XIX). All the octopuses were retrained at the same time of day and 24 hr after they were fed. In order to achieve this the groups were trained at different times of day and at different hunger levels. There were small differences in the training performances, but they did not correlate with the retraining performance. It was concluded that the variations in retention performance were a function of the retention interval. Both dips in the retention curve reappeared in a second experiment in which octopuses were trained with a 10-V a.c. shock to a criterion of no attack for 3 min and retrained at different times of day between 9.30 and 21.30 hr. With this higher shock voltage there were fewer attacks to criterion, resulting in a coarser measure of retention, and only the second dip was significant.

If the amount of training was increased or decreased, the overall level of retention performance could be raised or lowered, causing a partial or complete obliteration of the dips. Similarly, in his work with cuttlefish, Messenger (1973*a*) found that with the length of the initial training session reduced from 20 min to 3 min there was no difference in retention performance after intervals of 15, 30, 60, and 120 min. The results suggest that multiphasic retention curves can be obtained only within a limited set of training conditions. It has been suggested (Sanders and Barlow, 1971; Messenger, 1973*a*) that these curves may reflect the time courses of more than one memory system, with the initial fall in retention performance resulting from the loss of information from a shorter-term memory before the longer-term memory is available to control behavior. Multiphasic retention curves that appear to be comparable with those obtained with cephalopods have also been reported for fish, birds, and mammals (for references see Sanders and Barlow, 1971; and Barlow and Sanders, 1974).

Kovačević and Rakić (1971) have reported circadian variations in the performance of octopuses on a visual discrimination between complex figures. As yet it is not known if such changes can be obtained with the type of passive-avoidance training used in the experiments described above. The fact that Sanders and Barlow (1971) obtained essentially the same retention curves in two experiments in which retraining was either restricted to one time of day or occurred throughout the day suggests that circadian fluctuations are not a major variable in this situation. As an alternative explanation of multiphasic curves, however, this possibility warrants further investigation.

Boycott and Young (1955*b*) have demonstrated that a discriminant response, well established when octopuses are trained over a period of several days not to attack a crab shown with a white square, will survive deep anesthesia with 2–4% urethane, faradic stimulation of the vertical lobe, and up to 10 sec of stimulation with a 150-V a.c. current applied through saline pads to the head. When the octopuses recovered from these treatments they attacked crabs shown alone but not crabs shown with the white square.

If octopuses that have been trained to attack a vertical white rectangle are given 30 trials during which attacks on the rectangle are punished, they rapidly learn not to attack and retention is good 2 days later (Maldonado, 1968). Electroconvulsive shock (ECS; 12 V 400 mA for 2 sec) given 1 min after the 30th trial resulted in significantly more attacks during the retention tests given 2 days later (Table XX). Five octopuses trained on a brightness discrimination for ten sessions of 16 trials over a period of five days and given a similar ECS treatment on day 6 subsequently showed perfect retention (Young, unpublished data). Using the shuttlebox training procedure described in section V, C, Maldonado (1968) has shown that

Table XX. The Effect of Electroconvulsive Shock on the Retention of Training Not to Attack a Vertical Rectangle

(Data from Maldonado, 1968)

		Mean number of attacks		
		Training trials		Retention test trials
Treatment	N	1–15	16–30	1–15
Control group	15	2.80	1.00	1.80
ECS group	16	2.75	0.75	5.80

Table XXI. The Effect of the Length of the Training–ECS Interval on the Retention Performance of Octopuses Trained in a Shuttlebox—Retention Measured in Terms of a Decrease in the Number of "Incorrect Crossings," i.e., the Number of Times that an Octopus Returned from the Safe Light Box to the Dark Box, in which Shocks Had Been Received

(Data from Maldonado, 1969)

		Mean number of incorrect crossings		
Treatment	N	Training	Retraining	Probability
No ECS	55	3.17	1.48	<0.001
ECS 1 min	50	2.65	2.56	>0.05
ECS 6 hr	50	3.77	1.48	<0.005

ECS given within 1 min of the termination of the 20 training trials causes significant deficits in the retention performance measured three days later. A further experiment (Maldonado, 1969) confirmed this result and also demonstrated that if the ECS treatment was given 6 hr after training the retention performance was similar to that of a control group not given ECS (Table XXI). These are the first demonstrations in an invertebrate of the retrograde amnesia effect of ECS on recent learning.

VII. BRAIN LESIONS AND LEARNING

A. Separate Centers for Visual and Tactile Learning

The experiments to be considered here fall into two main categories: those designed to locate an area of the brain that is essential for a particular type of learning and those that attempt to ascertain the function of an individual lobe of the brain. As a result of these studies it has become clear that the centers for visual and tactile learning are, to a very great extent, separate in the brains of cephalopods (Fig. 16a). Lesions in the tactile system of octopuses do not disrupt visual learning, and touch learning can still proceed after complete removal of the visual system (see section VII, C). Further evidence for the separate location of the tactile center comes from comparative studies. The pelagic octopod *Argonauta* (Fig. 16b) has ceased using its arms as distance receptors and the tactile lobes have become secondarily reduced (Young, 1965*c*). In the cuttlefish the arms are merely used to manipulate the prey, and here the tactile lobes are found to be poorly developed compared with those of octopuses (Fig. 16c). Young (1963*c*, 1971) described the paired visual and tactile centers and discussed the relationship between the learning systems and the simple reflex biting response.

So many of the experiments employing ablation techniques have investigated the effects of lesions to the vertical lobe that these data will be considered separately after the other studies of the visual and tactile areas of the brain.

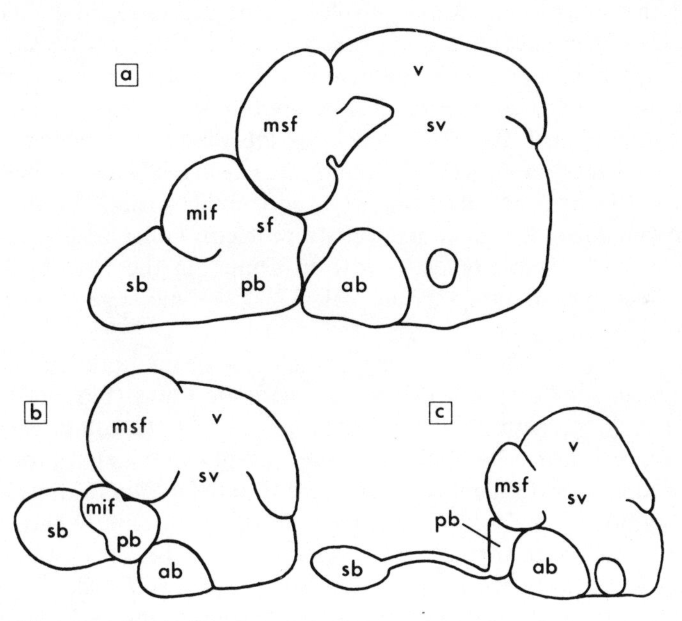

Fig. 16. Diagrams of sagittal sections through the supraesophageal lobes of (a) *Octopus*, (b) *Argonauta*, and (c) *Sepia*. In *Octopus* the median superior frontal, vertical, subvertical, and optic lobes form the visual system, while the tactile system consists of the median inferior frontal, lateral inferior frontal, subfrontal, and posterior buccal lobes. The lateral superior and inferior frontal lobes and the optic lobes are not visible in this section. Note the reduction of the tactile centers in *Argonauta* and their poor development in *Sepia*. *ab*, anterior basal; *msf*, median superior frontal; *mif*, median inferior frontal; *pb*, posterior buccal; *sb*, superior buccal; *sf*, subfrontal; *sv*, subvertical; *v*, vertical.

B. The Visual System

Boycott and Young (1950, 1955*b*, 1957) demonstrated the involvement of the median superior frontal, vertical, and optic lobes in visual learning. The organization of the visual system has been described by Young (1964*a*, *b*). After removal of the median superior frontal and vertical lobes, visual learning is still possible but greatly impaired. This finding, together with the discovery (Young, 1960*c*) of neurons in the optic lobes with elongated dendritic fields, predominantly with horizontal and vertical orientations, led to the suggestion that visual memories are located in the optic lobes (Young, 1965*b,c*). This theory is difficult to test because extensive lesions to an optic lobe, or its removal, result in blindness on the operated side; however, three studies have implicated the optic lobe in visual learning. Boycott (1954) reports that the habituation of avoidance responses occurs in octopuses with all of the supraesophageal lobes of the brain removed, indicating that the neural change involved must be restricted to the optic lobes and/or their subesophageal connections. Muntz (1963) has shown the importance of the optic lobes by demonstrating that intraretinal transfer will not occur if the connections in the optic lobe between the projection areas of the trained and tested parts of the retina are severed. Parriss (1963) found that, compared with controls, octopuses with lesioned optic lobes showed impaired relearning of a horizontal/vertical discrimination after interpolation of a diamond/square problem. Other lesion studies of the visual system have tended either to center on the functions of the vertical lobe system (see section VII, E) or to investigate transfer of training.

The paired optic lobes lie one on each side of the brain and are connected to the supraesophageal lobes by the optic tracts (Fig. 4). The two lobes are in direct contact through the large ventral optic commissure (Fig. 17a) and possibly also through the smaller dorsal commissure. Fibers from the optic lobes pass to the median superior frontal lobe via the lateral superior frontal lobes. Output from the median superior frontal lobe is via the superior frontal-vertical tract to the vertical lobe. Fibers from the vertical lobe pass to the subvertical lobe, from which axons run back to the optic lobes. There is much transverse crossing of fibers in the superior frontal-vertical tract and in the vertical lobe itself, through which the two optic lobes may be indirectly connected.

Interocular transfer of training does not occur when the direct connections between the optic lobes, which run in the optic commissures, are severed. The role of the indirect pathway via the vertical lobe has been studied by Muntz (1961*b*, *c*). He trained octopuses monocularly not to attack a crab and tested for transfer by presenting the crab to the untrained eye (see section VI, C). The performances of three groups of octopuses were compared. Group 1 had the superior frontal-vertical tract bisected in the midline by a vertical cut in an anterior–posterior direction (Fig. 17A). Group 2 had small control lesions in the same plane but at the posterior end of the vertical lobe beyond the furthest extent of the tract. Group 3 had varying amounts of the vertical lobe removed. In group 3 interocular transfer was inversely proportional to the amount of the vertical lobe removed; complete removal prevented transfer. There was no transfer in group 1 animals with the superior frontal-vertical tract bisected, while the octopuses with control lesions at the back of the vertical lobe showed perfect interocular transfer (Table XXII). In a second experiment, Muntz (1961*c*) obtained a similar result using a vertical/horizontal rectangle discrimination. When the superior frontal-vertical tract was bisected before training there was no transfer, but tests given to octopuses bisected after training showed that transfer had occurred (groups 2 and 3 in Table XXIII).

As bisecting the tract is known to prevent interocular transfer of learned behaviors, it is reasonable to assume that with the superior frontal-vertical tract bisected the untrained eye does not have access to information stored on the contralateral side of the brain. If so, the results obtained with groups 2 and 3 indicate that the engram is bilaterally located. Groups 4 and 5 (Table XXIII) were originally included to determine the involvement of the optic lobes in the storage of visual memories, but their usefulness was dependent on unilateral location of the engram. It can be noted, however, that equivalent performances were obtained with octopuses from which the optic lobe of either the trained or the untrained side had been removed after monocular training. This result confirms the finding of Boycott and Young (1955*b*) that after monocular training on a visual discrimination removal of the optic lobe from the trained side does not disrupt performance. In summary, it appears that for interocular transfer of training to occur the

Table XXII. Interocular Transfer and Lesions to the Vertical Lobe System in Octopuses Trained Not to Attack a Crab

(Data from Muntz, 1961*b*)

			Proportions of attacks/trials	
Group	Operation	*N*	Trained eye	Untrained eye
1	Bisected tract	9	3/39	38/38
2	Control	6	0/24	1/24
3	Vertical lobe removal	17	21/102	73/95

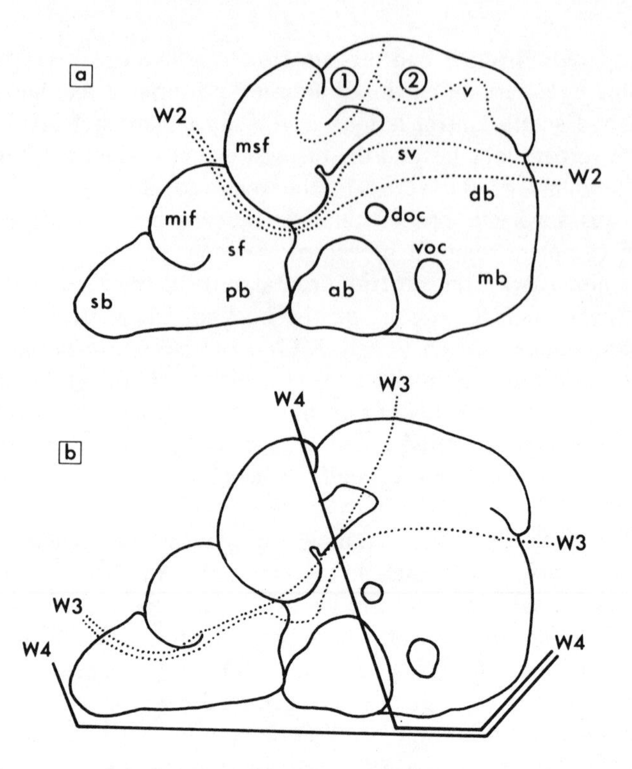

Fig. 17. The extents of the lesions used by Muntz and Wells in their studies of interocular transfer plotted on diagrams of sagittal sections of the supraesophageal lobes. All of the lesions were midline line vertical cuts in an anterior–posterior direction. a, *1*, the minimum extent of Muntz's experimental lesions in the median superior frontal-vertical tract that prevented interocular transfer; *2*, the maximum extent of control lesions in the rear of the vertical lobe that did not prevent transfer in this experiment, *W2*, the minimum and maximum extents of Wells's group 2 lesions that split the median superior frontal and vertical lobes *db*, dorsal basal; *doc,* dorsal optic commissure; *mb,* median; *voc*, ventral optic commissure; other abbreviations as in Fig. 16. b, *W3*, the minimum and maximum extents of Wells's group 3 lesions with the split including the inferior frontal lobe; *W4*, the minimum and maximum extents of Wells's group 4 lesions, which severed the optic commissures. (Data from Muntz, 1961*b*; Wells, 1970.)

transverse crossings in the median superior frontal-vertical lobe system must remain intact, as the optic commissures alone are insufficient.

A different picture of the neural basis for interocular transfer has been suggested by the detour experiments described in section VI, D. With practice in this situation the octopuses do not reduce the number of errors made, nor do they cease the futile attempts to attack the crab through the glass before detouring along the corridor. They do, however, appear to learn to continue along the corridor once they have entered it, because both the number of unfinished detours and the detour times decreased significantly with practice (Wells, 1970). In this paper Wells studied the effect on detour performance of partial or complete splitting of the supraesophageal lobes into equal halves by a median cut in an anterior–posterior direction. Group 1 were controls. In group 2 the median superior frontal and vertical lobes were split. In group 3 the split was extended to include the inferior frontal lobe. In group 4 the lesion divided the basal lobes and severed the optic commissures (Fig. 17). On detours that took longer than 30 sec to complete, there were no differences between the four groups in terms of per cent errors, whereas on faster detours, although the lesioned octopuses (groups 2–4) did not differ, they all made significantly more errors than the controls (Table XXIV).

Although the octopuses of group 4 (with the optic commissure severed) made no more errors than the others, they made markedly more abortive runs, i.e., detours which were begun, but not finished, within the 5-min period allowed. Octopuses from groups 1–3 made abortive runs on 27% or less of their trials, while group 4 animals failed to finish detours on 28% to 56% of their trials. The group 4 octopuses often vacillated as if commands from the two sides of the brain were in unresolved competition. This view was confirmed when three of these animals were tested again after uni-

Table XXIII. The Performance of Various Groups of Octopuses Trained Monocularly on a Horizontal/Vertical Visual Discrimination

(Data from Muntz, 1961*c*)

			Percent correct responses	
Group	Operation	*N*	Trained eye	Untrained eye
1	None	4	81.3	87.5
2	Tract bisected before training	5	84.4	50.6
3[a]	Tract bisected after training	5	74.4	63.1
4	Optical lobe removed from trained side	5		72.5
5	Optic lobe removed from untrained side	4	72.6	

[a] Scores for the untrained eye are the result of testing, while those of the trained eye arise from further training. During the four sessions contributing to these scores, the performance with the trained eye improved, while that with the untrained eye showed extinction. The performances on the first session were, however, identical: 72.5% correct for each eye.

Table XXIV. The Effect of Splitting Parts of the Supraesophageal Brain on the Performance of Octopuses on a Delayed Response in a Detour Task

(Data from Wells, 1970)

Group	Lesion (see Fig. 17)	Percentage of errors with the number of trials on which it is based		
		Delay 0–30 sec	Delay 31–60 sec	Delay >60 sec
1	Controls	9% (201)	27% (151)	45% (194)
2	SF and V split	17% (82)	27% (66)	39% (99)
3	SF, V, and IF split	19% (32)	27% (33)	46% (69)
4	Optic commissure severed	19% (97)	24% (111)	—

SF, superior frontal; V, vertical; IF, inferior frontal lobe (see Fig. 17).

lateral blinding and all showed considerable reductions in the numbers of trials on which abortive runs occurred.

Octopuses use only one eye to fixate the crab and to lead the subsequent detour, but the leading eye is sometimes changed as the animal enters the corridor. On these eye-change trials the eye that leads during the detour has not seen the position of the crab. The occurrence of competing responses from the two sides of the brain could account for the lower number of completed detours that are recorded on eye-change trials by oc-

Table XXV. The Proportion of Completed Detours Occurring on Eye-Change Trials for Octopuses with Different Parts of the Supraesophageal Lobes Split

(Data from Wells, 1970)

	Completed detours	Eye-change detours	Percent
Controls	698	254	37
Superior frontal and vertical lobes split	345	106	31
Optic commissure cut	209	25	12

topuses with the optic commissures cut (Table XXV). Detours are more likely to be completed by these animals when the eye that sees the crab leads throughout the run. It appears that the optic commissures, or structures closely associated with them, enable the exchange of information necessary to prevent conflict between the two sides of the brain and that the transverse crossings in the vertical lobe system are not required for this type of transfer.

If the detour is controlled by vision alone, equivalent performance on eye-change and no-eye-change trials would indicate that interocular transfer has occurred. The finding (Wells, 1964*c*) that unilaterally blinded octopuses make systematic errors to the unblinded side (see section VI, D) suggests that the detour is visually controlled. Two experiments have considered the possibility that interocular transfer occurs on eye-change trials. Wells (1967) compared the performances of octopuses with the vertical lobe removed with the performances of controls. Although the lesioned octopuses made more errors than the controls, both groups made the same proportion of errors on the eye-change and no-eye-change trials (Table XXVI). It would appear that destruction of the transverse crossings in the vertical lobe system does not prevent interocular transfer of the information required for a successful detour. Further analysis of data from octopuses with the optic commissure cut, however, suggests that this conclusion may not be justified. Although octopuses with the commissure cut make fewer completed detours on eye-change trials than the controls, when they do run they are at least as successful (Table XXVII). Given this surprising result there are three possible conclusions: (1) assuming the detour to be guided by vision alone, that interocular transfer of recently acquired information occurs via the subesophageal lobes; (2) that the detours

Table XXVI. The Performances of Octopuses, with and without the Vertical Lobe Removed, on a Delayed-Response Task in a Detour Situation

(Data from Wells, 1967)

Type of trial	Type of octopus	Response		Percent errors
		Correct	Incorrect	
Eye-change	Control	50	4	8
	No vertical lobe	29	7	24
No-eye-change	Control	317	24	8
	No vertical lobe	197	42	21

may be predominantly visually guided, particularly in intact octopuses, but that other mechanisms are available and may be employed when brain lesions prevent efficient use of visual guidance; or (3) that if an octopus begins a detour by entering from one side of the corridor, there is a greater probability that it will turn to the same side at the other end of the corridor. As the octopus inevitably begins a detour by attacking a crab shown behind the window to one side of the entrance, this would mean that even with octopuses capable of only a chance level of performance there would be a greater probability of correct rather than incorrect turns.

The first conclusion seems unlikely to be correct. Further analysis of data from 33 completed runs made on eye-change trials by six octopuses with the optic commissure cut suggests that the second and/or third conclusions may be correct. On the runs when these octopuses maintained contact with the wall separating them from the crab, they tended to turn correctly at the end of the corridor, whereas on the two trials during which contact was maintained with the other wall an incorrect response was made (Table XXVIII). Clearly, these octopuses maintained contact with the wall separating them from the crab on most of their runs and, in doing so, achieved a high success rate. Very probably, at least in octopuses with cut commissures, success is dependent on the maintenance of "correct" wall contact. There remain, therefore, two possible explanations of their high success rate: (1) that bodily orientation is used to guide the correct response, or (2) that after running along one side of the corridor an octopus is more likely to turn to that side. At present we cannot choose between these alternatives.

Table XXVII. The Performance of Octopuses with and without the Optic Commissures Cut on a Delayed-Response Task in a Detour Situation Calculated Separately for Eye-Change and No-Eye-Change Trials

(Data from Wells, Unpublished)

	Eye-change trials			No-eye-change trials		
	Errors	Runs	Percent errors	Errors	Runs	Percent errors
Controls	50	234	21	115	435	26
Optic commissure cut	4	25	16	45	159	28

Table XXVIII. The Detour Performance of Octopuses with the Optic Commissures Cut on Eye-Change Trials

(From Data Made Available by Dr. M. J. Wells)

Outcome of detour	Movement and position of octopus in corridor: Against wall on crab side	Along the center	Against wall on the other side	A combination
Correct turn	23	1	0	4
Incorrect turn	3	0	2	0

C. The Tactile System

Studying the effects of lesions in the visual system on touch learning, Wells and Wells (1957*c*) found that complete removal of the optic lobes did not disrupt the ability of octopuses to learn a rough/smooth tactile discrimination. Six animals without optic lobes made an average of 21 errors over 96 trials compared with a mean of 18 errors obtained with 13 octopuses blinded by the cutting of the optic nerves. Partial or complete removal of the vertical lobe resulted in the lesioned octopuses' making more errors than controls during tactile training, but this lesion did not prevent learning by touch. The effects of vertical lobe lesions will be discussed in more detail later.

The effects of more extensive lesions in the supraesophageal lobes has been reported by Wells (1959*a*). In one experiment octopuses were trained with trials at 3–5 min intervals not to take a Perspex cylinder by the punishment of takes with a 6-V a.c. shock. "Control" octopuses with no more extensive lesions than removal of the optic lobes made very few takes (errors) under these conditions. More errors were made after removal of the vertical lobe, but extension of the lesion to include the basal lobes produced no further deterioration in performance. In fact, the animals with both the vertical and basal lobes removed made fewer errors than those with just the vertical lobe removed (Table XXIX). The reason for this is not apparent. Lesions to the inferior frontal system do increase the learning deficit, and when this system is entirely removed, leaving only the buccal lobe of the supraesophageal mass intact, no discrimination learning occurs. The extents of these lesions are shown in Figure 18.

Octopuses with similar lesions were trained on a rough/smooth discrimination task with trials given at intervals of 5 min in blocks of 20 (Wells, 1959*a*). The results obtained confirmed the conclusions drawn from

Table XXIX. The Performances of Octopuses with Lesions to the Supraesophageal Lobes of Various Extents When Trained Not to Take a Perspex Cylinder

(Data from Wells, 1959*a*)

Type of octopus	Controls	Vertical lobe (V) removed	V+ basal lobes (B) removed	V + B removed + damaged inferior frontal	All removed except buccal lobe
N	7	5	5	12	4
Mean takes (errors) over the first seven trials	2.4	4.6	2.5	5.7	7.0[a]
Range	2–4	1–7	1–5	1–7	—

[a] A further seven trials with these four octopuses resulted in 27 takes.

the previous experiment. The amount of training required to reach a criterion was increased by removal of the vertical lobes, but removal of the basal lobes produced no further deficit (Table XXX). Two octopuses with extensive damage to the inferior frontal system and another with this system completely removed failed to learn the tactile discrimination in 200 trials. Octopuses in which the lesions did not extend to this system (vertical

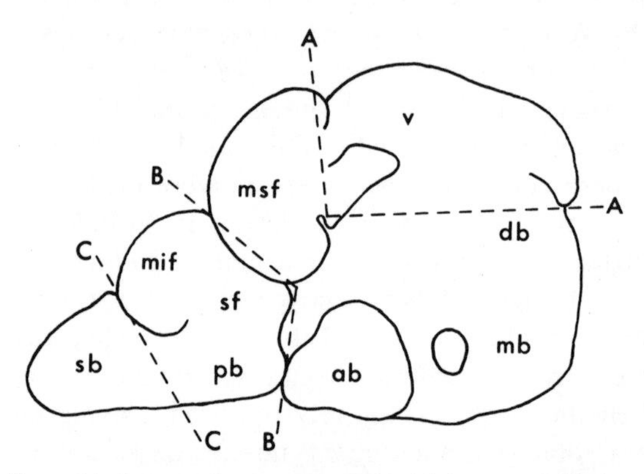

Fig. 18. The effect of lesions to the supraesophageal lobes on tactile learning. Removal of the vertical lobe (lesion *A*) caused an increase in the number of errors compared with controls, but extension of the lesion to include the basal lobes (lesion *B*) produced no further increase in errors. More extensive lesions, which damaged the inferior frontal system, produced a further increase in errors, while with complete removal of the inferior frontal system (lesion *C*), no evidence of learning was found. Abbreviations as in Fig. 3. (Data from Wells, 1959*a*.)

Table XXX. The Performances of Octopuses with Lesions to the Supraesophageal Lobes of Various Extents Measured in Terms of the Mean Number of Blocks of 20 Training Trials Required to Reach the Stated Criterion on a Rough/Smooth Tactile Discrimination Task

(Data from Wells, 1959*a*)

	Controls	Vertical lobe removed	Vertical and basal lobes removed
N	26	13	4
75% correct	2.35	4.00	4.25
85% correct	3.46	5.60 ($N = 10$)	5.75

and basal lobes only removed) reached an 85% correct criterion within 200 trials with a mean performance of less than 120 trials (Table XXX). The results obtained with octopuses with parts of the inferior frontal system damaged suggested that touch learning could occur in the absence of the median inferior frontal lobe, provided that parts of the lateral inferior frontal and subfrontal lobes remained intact. Evidence from the performances of a few animals suggested that transfer of training between the arms was impossible in the absence of the median inferior frontal lobe.

Wells and Young (1972) have confirmed the importance of the median inferior frontal lobe for tactile discrimination learning and have shown that

Table XXXI. The Performances of Octopuses with Lesions in the Vertical or Inferior Frontal Lobes When Given Four Trials per Day of Training Not to Attack Crabs

(Data from Wells, 1961*b*)

		Mean attacks (errors) per day		
Type of octopus	*N*	Day 1	Day 2	Day 3
Controls	9	4.0	0.9	0.8
Vertical lobe removed	8	5.0	4.5	4.1
Inferior frontal system lesioned	8	4.3	0.6	0.6

its role appears to be similar to that of the vertical lobe in visual learning (see section VII, E). Although touch learning can occur after removal of this lobe, it is severely impaired, particularly when the animal is trained to take the smooth and to reject the now-preferred rough stimulus. (Removal of the median inferior frontal lobe results in a change of preference from smooth to rough.) Despite the performance deficit produced by this lesion, octopuses that were trained on a rough/smooth discrimination continued to discriminate after the operation, albeit at a much lower level of accuracy. Retention of a tactile discrimination clearly survives removal of the median inferior frontal lobe. Wells and Young (1966, 1969*b*) have shown that splitting this lobe in the midline greatly disrupts transfer of the effects of tactile training; however, some transfer does occur after the commissural fibers in the median inferior frontal lobe have been severed, perhaps via the median superior frontal-vertical lobe system.

Although lesions to the inferior frontal lobe system have marked effects on tactile learning, they do not interfere with the learning of visual tasks. Boycott and Young (1955*b*) report that the performances of two octopuses trained to discriminate a crab shown alone from a crab presented with a white square were not affected by removal of the median inferior frontal lobe. Wells (1961*b*) trained octopuses with the lesions in either the tactile or the visual centers not to attack a crab presented at a distance. Controls and those with lesions in the inferior frontal system readily learned this visual task, whereas octopuses without the vertical lobe were severely impaired (Table XXXI). Wells (1961*b*) also trained three octopuses to discriminate horizontal/vertical rectangles after removal of the median inferior frontal lobe. Over the first six days of training with eight trials per day the three octopuses made 40 correct and 20 incorrect attacks compared with 47 and 17 by three controls. During a further six days' training the lesioned octopuses made 48 correct and only 7 incorrect attacks. These experiments provide further evidence that the tactile center is not involved in the learning of visual tasks.

Wells and Young (1965) began an extensive study of the tactile centers using split-brain preparations in which the supraesophageal lobes were divided into two halves by a median longitudinal cut. In some animals the lesion was found to lie to one side of the midline, causing varying degrees of damage on one side of the brain while leaving the other intact. The octopuses were trained on a rough/smooth tactile task using the arms of one side and either tested on the other or trained with the reverse problem. With the training restricted to one side no evidence was seen of transfer to the other, and one split-brain octopus was trained to make opposite responses on the two sides. Study of octopuses with various amounts of the tactile system destroyed showed that the ability to learn a tactile discrimination depended on the state of the subfrontal lobe and, in particular, on the presence of at least 10,000 of the small amacrine cells on the trained side of the split-brain (Table XXXII). Intact octopuses have about 2.5×10^6 of these amacrine cells on each side of the midline on the boundary between the subfrontal and posterior buccal lobes. Further evidence was obtained when three split-brain octopuses were trained to take a smooth cylinder presented to one side and to reject the same object when presented to the other side. One octopus with 84,000 amacrine cells on one side and 41,000 on the other learned this task making 68 takes on the positive side compared with 32 on the negative. In contrast, two octopuses with all the amacrine cells destroyed made a total of 106 positive and 101 negative takes.

Turning their attention to the problem of bilateral transfer of training, Wells and Young (1966) confirmed that with the supraesophageal lobe completely split in the midline no transfer of tactile training occurs from the trained to the untrained side. (All of the animals in this experiment were trained smooth-positive/rough-negative because the early experiments [Wells and Young, 1965] had revealed a preference for rough in split-brain octopuses [cf. Wells and Young, 1968*a*].) Ten split-brain octopuses given 160 training trials scored 68–96% correct responses (median 83%) with the trained side over the last 80 trials, while simultaneous tests on the untrained side resulted in scores ranging from 33% to 68% correct responses

Table XXXII. A Summary of the Relationship Between the Percent Correct Responses on a Rough/Smooth Tactile Discrimination and the Degree of Damage to the Subfrontal Lobe Obtained Using Split-Brain Octopuses

(Data from Wells and Young, 1965)

Condition of subfrontal lobe	Percent correct responses: Intact		10,000–45,000 amacrines intact		5,000 amacrines intact	
Rough +ve/smooth −ve	71, 68, 63, 63	66%	80, 74, 74	76%	70, 60, 60, 60, 55, 54, 53, 50	58%
Smooth +ve/rough −ve	80, 76, 71	76%			60, 46	53%

(median 50%) over 80 test trials. A further eight octopuses were given 80 training trials before the supraesophageal lobes were split, followed by another 80 trials after the operation, interspersed with 80 test trials to the untrained side. Unfortunately, in only two of these eight octopuses was the split exactly in the midline, causing little or no damage to either side, but both of these animals showed significant transfer, although performance in the unreinforced transfer tests (66% and 61% correct) was lower than that of the trained side during its postoperational training (88% and 79% correct, respectively). As transfer does not occur after splitting, it appears that with training restricted to one side the engram for a tactile task is still bilaterally located in intact octopuses.

Recently, Young (personal communication) obtained acquisition and extinction data from 30 split-brain octopuses with extensive lesions in the tactile centers of one side of the brain. The lesions were of three types: (1) total removal of the median inferior frontal and superior buccal lobes; (2) total removal of the median inferior frontal lobe, leaving the superior buccal lobe intact; and (3) complete severance of the cerebrobrachial tract, cutting off the input to the median inferior frontal lobe from the subesophageal lobes and arms. All three lesions, if performed bilaterally, would be expected to prevent tactile discrimination learning, yet the lesioned sides showed a better-than-chance level of response when both sides were trained on a smooth-positive/rough-negative tactile discrimination. In addition, there was a positive correlation between the performances of the lesioned and intact sides of the same octopus. Nine split-brain octopuses with these extensive lesions on both sides showed no such low-level discriminant responding. In the absence of other explanations, these findings suggest either that the arms of one side have limited access to information stored in the contralateral superesophageal lobes or that tactile information is stored outside the superesophageal lobes in subesophageal centers or in the arms. Further study is required to establish this phenomenon and to decide between these alternative explanations.

D. The Vertical Lobe System and Learning in the Cuttlefish

The involvement of the vertical lobe in learning and memory was first shown by the experiments of Sanders and Young (1940) with the cuttlefish. When a prawn is placed in the visual field of a cuttlefish, the latter shows first attention, then approach, and finally attack. If, as the cuttlefish approaches, the prawn is drawn back behind an obstacle, normal cuttlefish will continue to approach and follow the prawn round the obstacle, i.e., they will hunt a prey that has passed out of sight. Cuttlefish with the

Table XXXIII. The Effect of Vertical Lobe Removal on the Ability of Four Cuttlefish to Hunt for a Prawn That Has Passed Out of Sight

(Data from Sanders and Young, 1940)

	Trials	No response	Attention only	Attention and approach	Hunting
Before operation	13	2	—	—	11
After operation	62	6	17	36	3(?)

vertical lobe removed never showed hunting. The data obtained from four cuttlefish tested before and after vertical lobe removal are given in Table XXXIII. The three possible instances of hunting after vertical lobe removal were all obtained from one cuttlefish that approached very rapidly and was probably carried beyond the obstruction by its own momentum. During the other ten trials with this cuttlefish it showed attention and approach only.

On hatching from their large, yolky egg, young cuttlefish look and behave much like adults although they attack and feed on *Mysis,* a small crustacean, rather than prawns. In very young cuttlefish the vertical and median superior frontal lobes are relatively small and poorly developed (Wirz, 1954), but they develop and increase in size more rapidly than the rest of the brain during the first few months of life. Two reports have correlated these neuroanatomical changes with the ability of young cuttlefish to learn. Wells (1962*b*) showed that newly hatched cuttlefish were apparently unable to learn not to attack prey presented behind glass. A fall in the number of attacks did occur, but this was attributed to fatigue as these young animals settled down to a steady rate of attacking that, it was suggested, may be determined by the rate of mobilization of food reserves. Adult cuttlefish rapidly stop attacking in this situation (see section V, C). Five young cuttlefish, two 3 weeks, two 4 weeks, and one 5 weeks old, were given the same training with the addition of a 5–6-V shock for each attack. Only the five-week-old and one four-week-old animal learned not to attack. Messenger (1973*b*) studied learning not to attack *Mysis* presented behind glass in cuttlefish at 7, 28, 56, and 112 days after hatching. Eight 3-min presentations were made at 30-min intervals. Curves of errors against trials became steeper with age, and the 28-, 56-, and 112-day groups showed a significant reduction in attacks from the first to the eighth presentation. In a test given 24 hr after the eighth presentation, only the 112-day group showed significant retention. The relative sizes of the median superior frontal and vertical lobes, as seen in a median sagittal section, was found to parallel the improvement in learning. At 112 days these lobes are equivalent to those of an adult cuttlefish and the rate of learning is virtually identical (Messenger, 1973*a*).

E. The Vertical Lobe and Visual Discrimination Learning in Octopuses

The effects on behavior of lesions to the vertical lobe system have been extensively studied by Young and his associates. The general behavior is not greatly altered, although octopuses without vertical lobes tend to wander about the tank more than normal, unoperated controls (Young, 1961*a*). Untrained octopuses show altered tendencies to attack the plastic shapes that are used for visual discrimination training. Animals that attacked infrequently before the operation tended to attack more frequently afterwards and vice versa (Young, 1958*a*). These changes are not seen after a dummy control operation (Table XXXIV). Similar changes are seen on preference tests. When tested with a black and a white disc in cream-colored tanks, nine octopuses that made 72 attacks on the white and 96 on the black before removal of the vertical lobe made 115 attacks on the white and 86 on the black after the operation (Young, 1968). Further evidence of the preference reversal was obtained from the results of discrimination training with these shapes. Comparable changes have been noted in studies of discrimination by touch (see section V, E).

Young (1959) investigated the effect of vertical lobe removal on the performance of octopuses over a series of unreinforced trials. When plastic shapes that had not previously been associated with food were used, the

Table XXXIV. The Number of Attacks Made on a White Vertical Rectangle before and after a Dummy Control Operation and Removal of the Vertical Lobe

(Data from Young, 1958*a*)

		Total attacks on a white vertical rectangle								
Controls	Before	1	3	4	6	8	11	17	17	
	After	2	0	8	5	11	2	36	12	
	Difference	+1	−3	+4	−1	+3	−9	+19	−5	
No verticals	Before	3	4	6	6	6	8	11	32	33
	After	13	19	11	18	23	36	24	13	6
	Difference	+10	+15	+5	+12	+17	+28	+13	−19	−27

Table XXXV. The Mean Percent Correct Responses Scored by Octopuses Trained with Trials at 40-min Intervals on a Visual Discrimination with a Horizontal Rectangle Positive and a Vertical Rectangle Negative before and after Removal of Various Amounts of the Vertical Lobe

(Data from Young, 1958*b*)

Preoperational training		Percent removed	Retention test 10 trials	Postoperational training	
Trials 1–20	Trials 21–60			Trials 1–20	Trials 21–60
75	87	35–75	76	80	80
—	—		—	57	71
72	85	80–100	52	54	58
—	—		—	44	56

decrease in the frequency of attacks occurred at the same rate in normal and lesioned octopuses: habituation of the visual attack response appears to be unaffected by vertical lobe removal. When shapes that the octopuses had learned to attack were used, however, the attack rate decreased more rapidly in octopuses without vertical lobes: extinction would appear to occur faster in the lesioned octopuses. It is possible, however, that the different extinction performances result from an initial difference in the strength of the original learning rather than a faster rate of extinction in the absence of the vertical lobe. Further work is required to elucidate this point.

Visual discrimination learning in octopuses is markedly affected by vertical lobe removal. When the training procedures employed a crab as part of the stimulus complex (e.g., crab alone versus crab plus white square), octopuses without the vertical lobe could not be trained, nor could retention of preoperational training be demonstrated (Boycott and Young, 1957). If more than 50% of the vertical lobe was removed, the octopuses showed a reduced learning ability that correlated with the amount of tissue removed. When crabs were excluded from the stimulus complex, somewhat different results were obtained. The performance of octopuses trained before the operation was impaired in retention tests given after removal of the vertical lobe, but definite evidence of retention of the visual discrimination was found (Young, 1958*b*). Table XXXV shows the results obtained from octopuses trained on a visual discrimination with eight trials per day at intervals of about 40 min. (In a few of these animals the superior frontal–vertical tract, the major input to the vertical lobe, had been cut, and the figure given for the extent of the lesion is the percentage of the

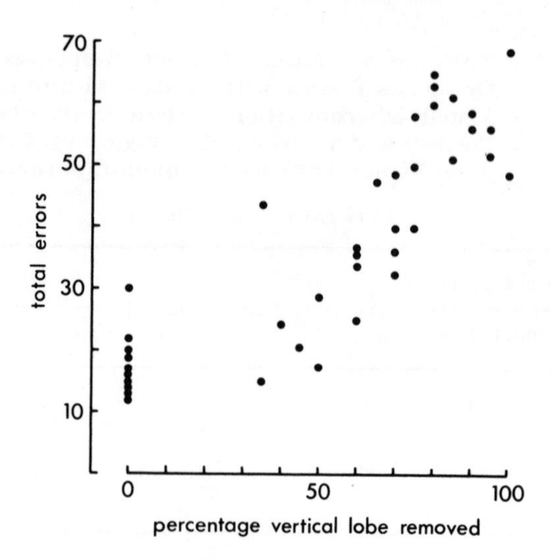

Fig. 19. The correlation between total errors made on a visual discrimination task and the percentage of vertical lobe removed. Note that with less than 50% removed performance is similar to that of controls. The octopuses were trained with a white horizontal positive and a white vertical rectangle negative for 60 trials at the rate of eight per day spaced at intervals of not less than 40 min. (Data from Young, 1958*b*.)

tract cut. Their performance was similar to octopuses with the same percentage of the vertical lobe removed.) Octopuses trained before the operation showed marked performance deficits after vertical lobe removal, but there was some retention, as comparison with the performance of octopuses trained only after the operation shows. The magnitude of the deficit is correlated with the amount of tissue removed (Fig. 19). Further tests using three octopuses with 85%, 95%, and 100% of the vertical lobe removed showed that their performance improved when unrewarded trials were given at 5-min intervals. Reintroduction of feeding produced increased attacks on the negative stimulus. In another experiment (Young, 1960*d*) in which the intertrial interval was 5 min rather than 40 min, similar deficits in visual discrimination learning were found in the octopuses without vertical lobes. With trials at short intervals the lesioned octopuses characteristically showed an improvement in performance during each training session, which largely disappeared by the start of the next one. However, a steady improvement over each session was apparent (Fig. 20) although it was markedly inferior to that of the controls. After 120 training trials the octopuses without vertical lobes were making three to four times as many errors as the controls. A comparable group of seven octopuses that were trained before and retrained after vertical lobe removal made significantly fewer errors over the last 80 retraining trials than did the octopuses trained only after vertical lobe removal (Fig. 20). A result such as this suggests that some of the effects of the prior training have survived removal of the vertical lobe. The reason for the initial poor retraining performance of this train-operate-retrain group is not readily apparent. Errors were not only the result of failure to attack the positive stimulus. The percent-correct-response score, which ignores failures to attack, was also initially lower for the train-operate-retrain group than for the operate-train group (Fig. 20). Perhaps the poor performance immediately after ventrical lobe removal is related to the tendency of this lesion to produce changes in previous preferences (see page 70).

The existence in octopuses of preferences for certain stimuli means that the difficulty of a particular discrimination often depends on which of the discriminanda is the positive. This is especially true for horizontal and vertical rectangles. The discrimination is easier when vertical, rather than horizontal, is positive. Using these two discriminations Young (1965*a*) has shown that the visual learning deficit that appears after vertical lobe re-

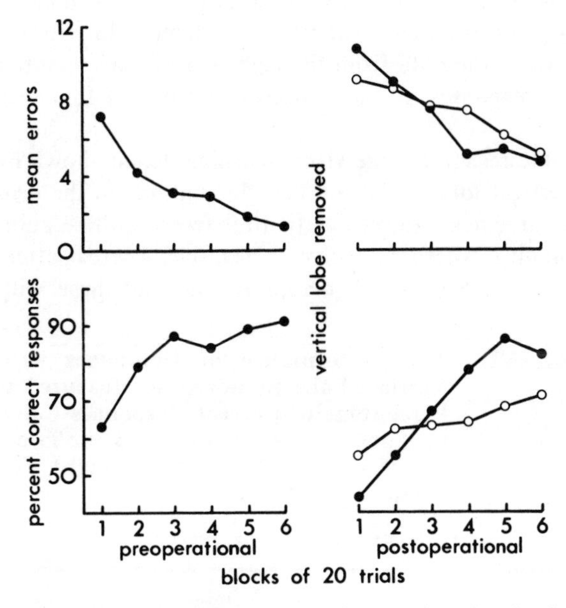

Fig. 20. The effect of vertical lobe removal on the retention of a visual discrimination as shown by a comparison of the performance of octopuses trained for the first time after the operation (open circles) with the postoperational retraining performance of octopuses that were initially trained before removal of the vertical lobe (filled circles). The performance, calculated as mean errors (upper graph) and mean percent correct responses (lower graph) is plotted for blocks of 20 trials (10 positive and 10 negative). In each group approximately half of the animals were trained horizontal rectangle positive/vertical negative and half the reverse. (Data from Young, 1960*d*).

Table XXXVI. The Performance of Octopuses with the Vertical Lobe Removed on an Orientation Discrimination as Measured by the Percentage of Correct Responses Made during Each Daily Session of 16 Trials

(Data from Young, 1965*a*)

Discrimination	Percent correct responses for each training session							
	1	2	3	4	5	6	7	8
Horizontal $^{+ve}$/vertical $^{-ve}$	48	49	46	45	51	49	51	56
Vertical $^{+ve}$/horizontal $^{-ve}$	63	61	69	79	72	75	75	73

moval is much more marked with the difficult than with the easy form of the task (Table XXXVI). The difference is even more marked when the improvement within training sessions is considered. This is shown in Table XXXVII, in which the data from the eight sessions are combined and the percent correct responses for each successive block of four trials is calculated.

One possible reason for the visual learning deficit shown by octopuses without the vertical lobe is the marked fluctuations in the level of attack caused by reinforcement. Immediately after feeding, these animals tend to attack any stimulus, whether positive or negative, whereas after shock they tend not to attack. Sometimes octopuses such as these will show dis-

Table XXXVII. The Performance of Octopuses with the Vertical Lobe Removed as Measured by the Percentage of Correct Responses Calculated for Each Successive Block of Four Trials with the Data from Each of the Eight Training Sessions Combined

(Data from Young, 1965*a*)

Discrimination task	Percent correct responses			
	Trials 1–4	5–8	9–12	13–16
Horizontal $^{+ve}$/Vertical $^{-ve}$	46	49	51	55
Vertical $^{+ve}$/Horizontal $^{-ve}$	60	69	72	75

criminant responses in the absence of reinforcement, while showing no evidence of discrimination learning during training with rewards and punishments (Young, 1958*b*). If this analysis is correct, then octopuses with the vertical lobe removed should perform better on a simultaneous discrimination task when variations in the level of attack cannot affect the discrimination performance. Muntz *et al.* (1962) trained normal octopuses and others with 80–93% of the vertical lobe removed for 12 sessions of four trials each on a simultaneous discrimination between black and white squares. Black was chosen as the positive stimulus, as all of the octopuses had exhibited an initial preference for whites. Both groups learned rapidly, but the initial improvement and subsequent asymptotic performance were both better in the unoperated octopuses (Fig. 21). After the twelfth session the vertical lobe was removed from the normal octopuses and their training continued for a further 12 sessions. Although there was a considerable postoperational performance deficit, their performance level was better than that of the naive lesioned animals, indicating that at least some of the previous training has survived removal of the vertical lobe. When other octopuses were given successive training on the same discrimination for 128 trials over eight sessions, those with the vertical lobe removed never performed significantly above chance level, whereas the control octopuses learned readily (Fig. 21). Similar results were obtained in two other simultaneous training experiments using different visual discrimination tasks.

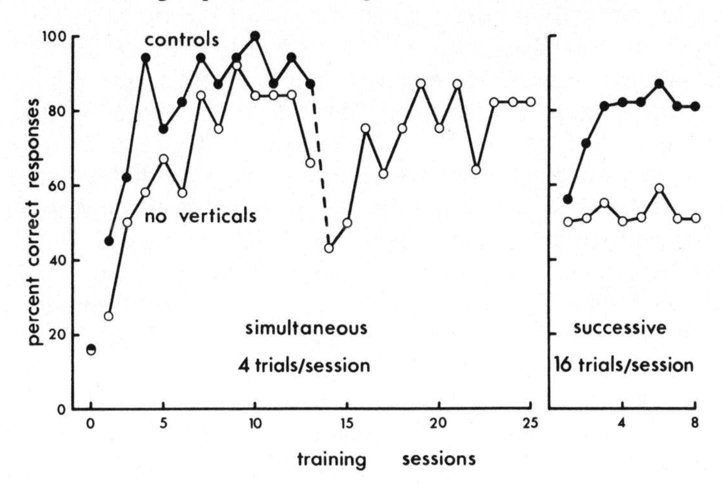

Fig. 21. A comparison of the performances of octopuses with and without the vertical lobe during training with both simultaneous and successive presentation of the stimuli. All of the octopuses were trained with a black square positive and a white square negative. The two points plotted for the simultaneous training session "zero" indicate the result of an initial preference test. After 12 sessions of simultaneous training, the vertical lobe was removed from the control octopuses and the training continued for another 12 sessions. (Data from Muntz *et al.*, 1962.)

We may now summarize the effects of vertical lobe removal on visual discrimination learning. When trials were given at intervals of about 1 hr and a crab formed part of the stimulus complex, little evidence of learning or of retention of preoperational learning was found. Subsequently, the use of a variety of modified procedures demonstrated (1) that octopuses with the vertical lobe removed could learn visual discriminations, albeit more slowly and to a lower asymptotic performance level than controls, and (2) that at least some preoperational learning survived the loss of this lobe. The successful modifications were:

a. Reducing the intertrial interval to 5 min (Young, 1958*b*, 1960*d*).
b. Presenting the discriminanda without crabs (Boycott and Young, 1957).
c. Withholding rewards and punishments (Young, 1958*b*).
d. Using simultaneous rather than successive presentation of the discriminanda, so that discrimination performance was not affected by a fluctuating level of attack (Muntz *et al.*, 1962).

The learning deficit is also less marked with easy discrimination tasks (Young, 1965*a*). These experiments suggested that the visual memory stores are situtated, at least to a great extent, outside the vertical lobe, and the optic lobes were proposed as a probable location (Young, 1961*a*; Muntz *et al.*, 1962).

Young (1960*d*, 1961*b*) has estimated that with successive presentations the learning of visual discrimination tasks is four to five times slower in octopuses with the vertical lobe removed than in controls. As we have seen (Fig. 19), the learning deficit is proportional to the amount of vertical lobe tissue removed, and with less than 50% removed performance is equivalent to that of unoperated octopuses. Other experiments have shown that the deficit is apparent both in learning not to attack and in learning to attack. When trained to attack a vertical white rectangle with eight trials per day, at intervals of either 5 or 50 min, a group of ten controls increased their performance from 65% attacks on day 1 to 95% on day 4, whereas ten octopuses with the vertical lobe removed remained at about 60% attacks for seven days (Young, 1961*b*). There were no marked differences between the two rates of training in this experiment. In a second experiment octopuses of both types were trained for four days at five trials per day to attack a white horizontal rectangle. The attacks of five control animals increased from 15% to 70%, whereas in the lesioned octopuses the increase was only from 10% to 20%. Training was now continued with a white vertical rectangle. The controls began at 80% and reached 90% plus in 15 trials, while the operated animals started at 40% and reach 75% in 20 trials. At the completion of 25 trials, the levels of attack had remained constant at 90%

and 75% respectively. The next day the octopuses were given three shocks in the presence of the white vertical rectangle, whether they attacked or not. Without further shocks the controls did not attack again for four days, while the lesioned octopuses began to attack again at the end of the first day on which they received the shocks. The progress of learning not to attack a stimulus is more readily demonstrated in training not to attack a crab. Octopuses without vertical lobes trained with trials at 4-min intervals showed a rapid drop in the number of attacks per trial, all having stopped attacking the crab after 8 trials, while others trained at 8-min intervals required 14 trials before they stopped (Young, 1965*a*). Controls trained at 8-min intervals stopped after 5 trials. Similar effects of vertical lobe removal on learning to attack and not to attack have also been reported by Maldonado (1965) and Young (1970).

Maldonado (1963 *a,b,c*) developed a detailed block diagram model representing the visual attack learning system of the octopus, in which some of the blocks were identified with specific lobes of the brain. The vertical lobe system was seen as an amplifier that generates the command to attack or retreat. The model, which incorporated much of the known data, including the sizes and numbers of nerve cells in each lobe and the connections between them, led to several predictions that were subsequently tested and confirmed.

F. The Vertical Lobe and Tactile Discrimination Learning in Octopuses

The effects of vertical lobe removal obtained with tactile training are similar but much less marked than the effects on visual learning. With simple tasks, such as habituation of the taking response and training with negative reinforcement not to take a particular stimulus object, similar acquisition rates are obtained from octopuses with and without the vertical lobe. Only a few presentations of an inedible Perspex cylinder at 2-min intervals are required before the octopuses reject the cylinder after only a cursory examination with an arm tip, whereas in earlier trials it is passed to the mouth for examination. Octopuses without the vertical lobe habituate in about the same number of trials as controls (the means were 4.8 and 4.7, respectively), although they characteristically spend longer examining the cylinder on the first trial: on average 14 min 19 sec as opposed to 2 min 23 sec (Wells and Wells, 1957*a*). If 6-V a.c. shocks are given each time the stimulus is taken, both control and lesioned octopuses learn in 4–5 trials (Wells, 1965*d*).

Clear differences appear when the results of training to discriminate between a smooth and a rough cylinder are considered. The effects of removing both the median superior frontal lobe and the vertical lobe, and of removing the median superior frontal lobe alone, were not distinguishable from simple vertical lobe removals, so the data have been considered together in Table XXXVIII. As well as requiring more training to reach the criterion of 75% correct responses than the controls (4.2 instead of 2.3 sessions), the lesioned octopuses also showed significantly greater variance in performance. These data were obtained with training at short intertrial intervals of 5 min. With longer intervals of about 1 hr and only 8 trials per day, the deficit shown by 12 lesioned octopuses was greater (Wells and Wells, 1957*c*). The difference between the performances of controls and operated octopuses was also increased when a more difficult tactile discrimination task was used. As in visual learning, the size of the deficit is proportional to the amount of the vertical lobe that is removed.

Wells (1965*d*) has analyzed the types of errors made by octopuses with lesions to the vertical lobe system during tactile discrimination training. He found that lesioned and control octopuses erred mainly by taking the negative stimulus. The pattern of errors was the same for both groups, with 70–80% of the errors being made on the negative trials. Once the octopuses had begun to discriminate, the proportion of negative errors remained stable at this figure despite great changes in the total number of errors made per session. Although the pattern of errors changed from the first to

Table XXXVIII. The Number of Sessions, Each of 10 Positive and 10 Negative Trials, Required to Train Lesioned and Control Octopuses to a Criterion of 75% Correct Responses in a Session Using a Rough/Smooth Tactile Discrimination Task

(Data from Wells, 1965*d*)

Type of octopus	Number of sessions of 20 trials required to reach 75% correct								
	1	2	3	4	5	6	7	8	9
V and MSF removed	3	2	4	3	1	0	2	2	0
V removed	0	0	6	2	2	1	1	1	1
MSF removed	1	1	0	0	1	1	0	0	0
All lesions	4	3	10	5	4	2	3	3	1
Controls	7	20	11	4	0	0	0	0	0

V, vertical lobe; MSF, median superior frontal lobe.

the second half of a training session, the change was the same for lesioned and control animals, with both making fewer negative errors. Hence, although octopuses with vertical lobe lesions make more errors than controls they do so by taking the negative stimulus and by failing to take the positive stimulus in proportions similar to those of controls.

When trained on a rough/smooth discrimination with trials at 1-hr intervals before and after vertical lobe removal, octopuses show considerable postoperational performance deficits (Fig. 22), but comparison with octopuses trained only after the operation indicates that at least some of the effects of tactile training survive vertical lobe removal (Wells and Wells, 1958*b*). With training at 5-min intertrial intervals, however, the effect of the operation is much less marked and the preoperational training clearly survives the loss of the vertical lobe (Fig. 22).

The retention of tactile discriminations has also been tested in octopuses trained for the first time after vertical lobe removal (Wells and Wells, 1958*a*). Controls and lesioned octopuses were trained to a criterion of either 75% or 85% correct and then overtrained for a similar number of trials. During overtraining the controls continued to improve faster, so their terminal training performance was slightly higher than that of the lesioned octopuses, but retention was similar in the two groups (Table XXXIX). In octopuses of both types there tended to be a swing toward

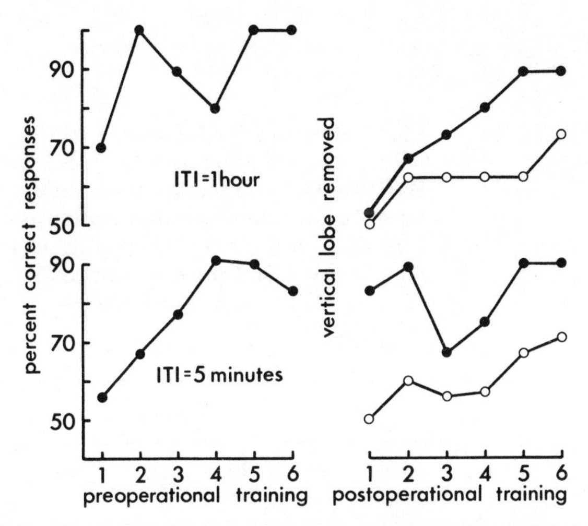

Fig. 22. The effect of vertical lobe removal on the retention of a rough/smooth tactile discrimination learned before the operation. Filled circles, octopuses trained before and retrained after removal of the vertical lobe, open circles, octopuses trained for the first time after vertical lobe removal. Note the marked effect of reducing the intertrial interval from 1 hr to 5 min. (Data from Wells and Wells, 1958*b*; Wells, 1962*a*.)

increased taking of the positive and the negative stimulus after the retention period. This swing was sufficient in some animals to mask retention of the ability to respond discriminantly. With these octopuses, however, discriminant responding could be demonstrated either (1) by repeated testing and so reduction of the level of taking by a process of extinction, or (2) by the giving of a small (6-V a.c.) shock for each take, no matter which stimulus was involved.

A study of tactile retention performance over much longer periods has been made by Sanders (1970*b*). At relatively short retention intervals of up to 30 days, the retention performance of octopuses without vertical lobes was of the order of that of controls. This finding is in agreement with that of Wells and Wells (1958*a*). At long intervals of 55 to 120 days, however, the retention performance of the lesioned octopuses was not only better than that of controls but also better than that of the lesioned animals tested after short retention periods (Table XL, Expt. I). Given the high scores at long retention intervals, it seemed improbable that the low scores obtained at short intervals by octopuses without the vertical lobe were a direct result of the length of time between training and retraining. Perhaps the interval between vertical lobe removal and retraining is the relevant variable, but in this experiment the two variables were confounded.

In a second study (Sanders, 1970*c*), short retention periods were combined with long operation/retraining intervals and high retention scores were obtained (Table XL, Expt. II). These results indicate that low retention scores are obtained from octopuses without vertical lobes only

Table XXXIX. The Retention Performance of Octopuses with and without the Vertical Lobe as Measured by the Mean Errors on a Rough/Smooth Tactile Discrimination Task after Training to a Criterion of 87% Correct—Scores for Octopuses Trained to 75% Correct Given in Parentheses

(Data from Wells and Wells, 1958*a*)

	Mean errors		
	Last 20 training trials	20 test trials after 5 days	20 test trials after 10 days
Controls	1.4 (2.3)	3.2	5.2 (6.8)
Vertical lobe removed	2.8 (2.9)	4.7	5.4 (7.4)

Table XL. The Effect of the Length of the Interval between Vertical Lobe Removal and Retraining on the Retention Performance of Octopuses—Number of Animals Contributing to Each Score Given in Parentheses

(Data from Sanders, 1970*b*,*c*)

Source	Retention period (days)		Retention score: Controls	Retention score: No vertical lobe	Operation/retraining interval (days)	
Expt. I	Short	10	0.82(7)	0.63(8)	18	Short
		20	0.76(7)	0.17(4)	28	
		30	0.40(12)	0.55(7)	48	
	Long	55	0.42(8)	0.97(3)	77	Long
		60	0.43(6)	0.87(2)	108	
		120	0.21(4)	0.94(4)	153	
Expt. II	Short	15	0.76(2)	1.03(4)	113	Long
		25	0.94(10)	0.87(6)	116	
		30	0.55(7)	—	149	

when they are retrained after short operation/retraining intervals, suggesting that during the period immediately following vertical lobe removal the lesioned octopuses experience a tactile performance deficit which disappears within 50–80 days. The lesioned octopuses trained in Expt. II were the first to be initially trained so long (91 and 98 days) after removal of the vertical lobe. Compared with controls these lesioned octopuses showed no tactile performance deficit. Evidence from two sources, initial training performance and retraining performance, supports the view that removal of the vertical lobe produces a relatively long-lasting, but nonetheless temporary, tactile performance deficit that disappears two to three months after the operation. Comparable experiments using visual discrimination training were not possible because of the severe acquisition deficit shown by octopuses without the vertical lobe. However, it is perhaps significant that the disappearance of the performance deficit was found with tactile training, as the vertical lobe forms a much more integral part of the visual than of the tactile learning system.

The retention scores obtained by the nine individual octopuses without vertical lobes that were retrained after the long retention intervals correlated with the amount of vertical lobe removed (Sanders, 1970*b*). This result was

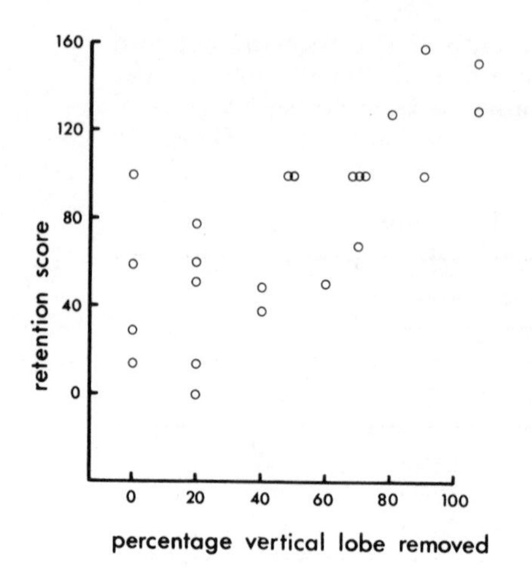

Fig. 23. The correlation between long-term tactile retention, as measured by retraining after 55–60 days, and the percentage of vertical lobe removed. (Data from Sanders, 1970*b*, *c*.)

confirmed by Sanders (1970*c*) with 18 octopuses, each trained against its rough/smooth preference. The data obtained in these two studies from animals retrained after 55–60 days are plotted in Fig. 23. Thus, it would appear that, whereas performance on a discrimination task soon after vertical lobe removal is *negatively* correlated with the amount of tissue removed (Wells and Wells, 1957*c;* Young, 1958*b*), retraining performance after long operation/retraining intervals is *positively* correlated with the percentage of vertical lobe removed.

G. Delayed Response, Delayed Reinforcement, and the Vertical Lobe

The studies described in section VI, E also investigated the effect of vertical lobe removal. The results of Wells and Young (1968*b*) suggest that the learning of both visual and tactile discrimination tasks is more seriously impaired by delays of reinforcement in octopuses with the vertical lobe removed than in normal animals. The appearance of strong preferences in the lesioned octopuses, particularly in tactile training, made assessment of the data difficult. At the longer delays, the octopuses without the vertical lobe tended to swing toward marked preferences, with some animals taking the positive stimulus but others taking the negative. However, with both tactile and visual discrimination tasks, it seems that the lesioned octopuses showed little evidence of learning when reinforcement was delayed 15 sec, while the controls learned with a 30-sec delay. Unfortunately, we cannot be certain that delayed reinforcement impairs the lesioned animals more than the controls because octopuses without the vertical lobe show a general discrimination learning deficit that may have produced the differences seen here.

A clearer answer to the effect of delay on the performance of these animals has been provided by the delayed-response studies. With enforced response delays of 10, 20, and 30 sec Dilly (1963) found no differences between the performances of control and lesioned octopuses (Table XLI). Wells (1967), using a detour through an opaque corridor to produce the delayed response, was able to measure performance after delays as long as 2 min. There was little difference between the controls and the octopuses without vertical lobes on short delays, confirming the results obtained by Dilly. As the delay lengthened, however, the proportion of errors made by the lesioned animals increased more rapidly than for the controls. At delays of over 1 min, octopuses without vertical lobes performed at chance level, whereas controls were still performing above chance level with delays of 2 min. This finding was confirmed by Sanders (1970*a*) when he repeated Dilly's experiment with longer delays (Fig. 24). Success on a delayed-response task implies efficient short-term storage of information. It appears that removal of the vertical lobe reduces the efficiency of this storage, perhaps by reducing the amount of information it can hold or by accelerating the rate at which information is lost.

In a second experiment (Sanders, 1970*a*), however, the performance of the lesioned octopuses equalled that of controls even at delays of 90 sec and 3 min (Fig. 24). The proportions of errors made by controls and octopuses without the vertical lobe were significantly different with delays of 60 sec in Expt. I but not with delays of 90 and 120 sec in Expt. II. The lesioned animals used in Expt. I had been operated on 27 days before testing, but those in Expt. II were tested 54 days after vertical lobe removal. This finding

Table XLI. The Performance of Control and Lesioned Octopuses with an Enforced Delayed Response of 30 sec

(Data from Dilly, 1963)

Type of octopus	Total responses	
	Correct	Incorrect
Controls	34	7
Vertical lobe removed	33	5
Medium superior frontal lobe removed	36	16

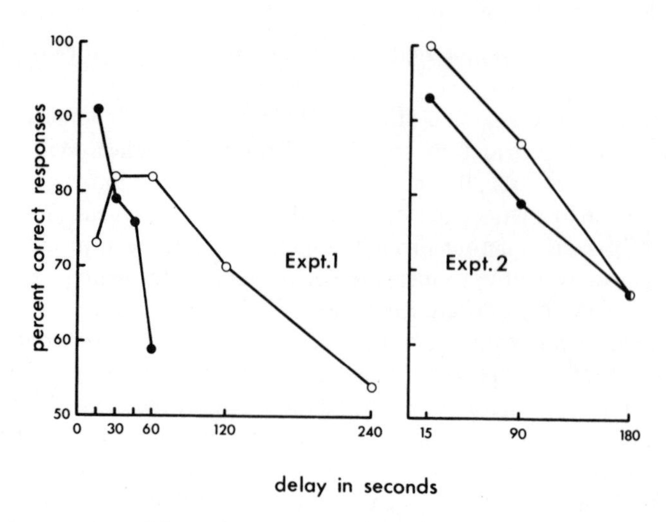

Fig. 24. A comparison of the performances on a delayed-response task of octopuses with and without the vertical lobe. Open circles, controls; filled circles, octopuses with the vertical lobe removed. The performance of the controls with a 15-sec delay in Expt. I is atypically low. Note the marked effect on the performance of the lesioned octopuses of the length of the interval between vertical removal and testing. This interval was 27 days for Expt. I and 54 days for Expt. II. (Data from Sanders, 1970*a*.)

suggests that the performance deficit seen in the delayed-response task may also be a temporary phenomenon that, like the tactile discrimination performance deficit, is apparent during the first weeks after vertical lobe removal but disappears about 2 months after the operation.

H. The Effect of Vertical Lobe Removal: A Summary

Removal of the vertical lobe has little or no effect on the general behavior of octopuses, but marked effects on behaviors that involve learning or memory are produced. The phenomena that have been observed in animals without the vertical lobe are enumerated here.

1. There is a reversal of preoperational levels of attack and preferences (Young, 1958*a*).
2. There is a marked impairment of the ability to learn both to attack and not to attack a stimulus (Young, 1961*b*, 1970; Maldonado, 1965).
3. There is a marked impairment of visual discrimination learning (Young, 1958*b*), which is four to five times slower in the absence of the vertical lobe (Young, 1960*d*, 1961*b*).

4. Performance is better with simultaneous rather than successive visual discrimination training (Muntz *et al.*, 1962).
5. The impairment is greater when crabs form part of the stimulus complex (Boycott and Young, 1957).
6. The impairment is greater for difficult discriminations (Young, 1965*a*).
7. The impairment is less when trials are given at short (5-min) rather than long (1-hr) intertrial intervals (Young, 1958*b*, 1960*d*).
8. Visual discrimination performance may improve when reinforcement is withheld (Young, 1958*b*).
9. The magnitude of the deficit is proportional to the percentage of the vertical lobe removed (Young, 1958*b*).
10. Although postoperational performance deficits are large, some evidence of the retention of preoperational visual discrimination training can be demonstrated after the vertical lobe has been removed (Young, 1958*b*, 1960*d*).
11. Habituation of the attack response to unfamiliar visual stimuli appears not to be affected by vertical lobe removal (Young, 1959).
12. Longitudinal bisection or removal of the vertical lobe prevents interocular transfer of information acquired during prolonged training (Muntz, 1961*b,c*).
13. Cuttlefish without the vertical lobe will attack prey but, should the prey move out of sight, will not follow it (Sanders and Young, 1940).
14. Learning ability in young cuttlefish improves as the vertical lobe system of the brain develops (Wells, 1962*b;* Messenger, 1973*b*).
15. Tactile habituation is unaffected by vertical lobe removal (Wells and Wells, 1957*a*).
16. Learning not to take a tactile stimulus is also unaffected (Wells, 1965*d*).
17. Tactile discrimination learning is slower in the absence of the vertical lobe (Wells and Wells, 1957*a;* Wells, 1965*d*).
18. The tactile discrimination learning deficit is less marked with short (5-min) than with long (1-hr) intertrial intervals (Wells and Wells, 1957*c*).
19. At 5-min intervals octopuses without the vertical lobe require about 50% more trials to reach a criterion (Wells and Wells, 1957*c*).
20. The size of the learning deficit is proportional to the amount of the vertical lobe removed (Wells and Wells, 1957*c*).
21. Lesioned octopuses make the same proportion of each type of error (positive and negative) as unoperated animals (Wells, 1965*d*).

22. Preoperational tactile training survives vertical lobe removal (Wells and Wells, 1958*b*).
23. Retention of a well-learned tactile discrimination is as good in lesioned as in normal octopuses over periods of a few days to a few weeks (Wells and Wells, 1958*b;* Sanders, 1970*b*) and appears to be even better over longer periods in the absence of the vertical lobe (Sanders, 1970*b,c*).
24. Retention performance at these longer intervals is proportional to the amount of vertical lobe removed (Sanders, 1970*c*).
25. Under conditions of delayed reinforcement octopuses without vertical lobes learn less well than controls (Wells and Young, 1968*b*).
26. Performance on a delayed-response task is impaired in octopuses without the vertical lobe (Wells, 1967; Sanders, 1970*a*).
27. The tactile discrimination learning and delayed-response performance deficits seen in octopuses during the first few weeks after vertical lobe removal appear to be relatively long-lasting, but nonetheless temporary, phenomena that disappear some two months after the operation (Sanders, 1970*a,b,c*).

I. Proposed Functions of the Vertical Lobe System

From the wealth of empirical data that have been collected over the last 20 years, it is clear that the vertical lobe and its circuits are intimately involved in the processes of learning and memory. Defining a specific role has, however, proved difficult. The many proposals that have been made may be roughly classified under eight headings, although some suggestions covered more than one of these functions.

Additional Memory Storage Units (Boycott and Young, 1957; Young, 1958*b*, 1960*d*, 1961*a,b*, 1962*b;* Wells, 1965*d*). The major support for this view is the marked performance deficit seen in the postoperational retraining performance of octopuses trained before vertical lobe removal.

Short-Term Memory System (Boycott and Young, 1955*b;* Young, 1958*b*, 1960*b*, 1970; Wells, 1967; Wells and Wells, 1957*c*; Wells and Young, 1969*a*). The observations that single reinforcements are effective for shorter periods in the absence of the vertical lobe; the improved acquisition performance obtained with trials at short, rather than long, intervals; and the inferior performance of octopuses without the vertical lobe on delayed response tasks and with delayed reinforcement support this view.

Readin Device (Boycott and Young, 1955*b*; Young, 1958*b*, 1960*b*, 1970; Wells, 1967; Wells and Wells, 1957*c*, 1958*a*). Many proposals have coupled the role of a short-term memory system with that of establishing the long-term memory.

Readout Device (Young, 1958*b*, 1961*a*, 1964*a*). Young has noted that the evidence for the vertical lobe's providing additional memory storage units could be equally well explained by this lobe's forming part of a readout device for memories located elsewhere.

Generalization and Transfer (Muntz, 1961*b,c*, 1962; Young, 1961*a*, 1962*b*). The optic commissures are insufficient for the interocular transfer of discrimination learning, for which the transverse crossings in the vertical lobe system are required. Generalization of learning to attack a rectangle will, however, occur in the absence of the vertical lobe.

Attack Inhibitor (Young 1958*b*, 1964*a,b*, 1965*c*; Maldonado 1963*b*; Nixon and Young, 1966). Young has proposed that the vertical lobe system serves to balance the response level, with the median superior frontal lobe promoting and the vertical lobe inhibiting attack. Evidence for this view comes from the tendency of attack levels and preferences to be reversed after lesions to this system. Differential effects of removing one or the other of the two lobes have proved hard to demonstrate. This difficulty is not surprising, as the entire output of the median superior frontal lobe passes to the vertical lobe.

Amplifier (Young, 1961*a*; Maldonado, 1963*b,c*, 1965; Nixon and Young, 1966). The vertical lobe is seen as a general amplifier both for information from receptors and the output of the memory stores. A particular role, that of increasing the value of the negative, "pain" signals from the nocireceptors so as to reduce the level of attack, is also suggested.

Motivation (Muntz *et al.*, 1962). After summarizing the evidence available at that time, these authors concluded that vertical lobe removal results in motivational difficulties and that the lobe normally serves to stabilize, and perhaps lower, the level of attack.

The most comprehensive proposals concerning the roles of the various lobes of the octopus brain in learning and memory are those suggested by Young (1964*a*, 1965*c*). He identifies two systems, each composed of four lobes arranged as an upper and a lower pair (Fig. 25A). The tactile system consists of the median and lateral inferior frontal lobes, beneath which lie the posterior buccal and subfrontal lobes. The corresponding lobes in the visual system are the median superior frontal and vertical lobes, plus the lateral superior frontal and subvertical lobes. In addition, the visual system includes the optic lobes, in which visual input is analyzed and the memory records are thought to be stored. Tactile memory records are thought to be located in the posterior buccal-subfrontal region. The upper and lower pairs of lobes in each system form two reexciting circuits in parallel. The role of

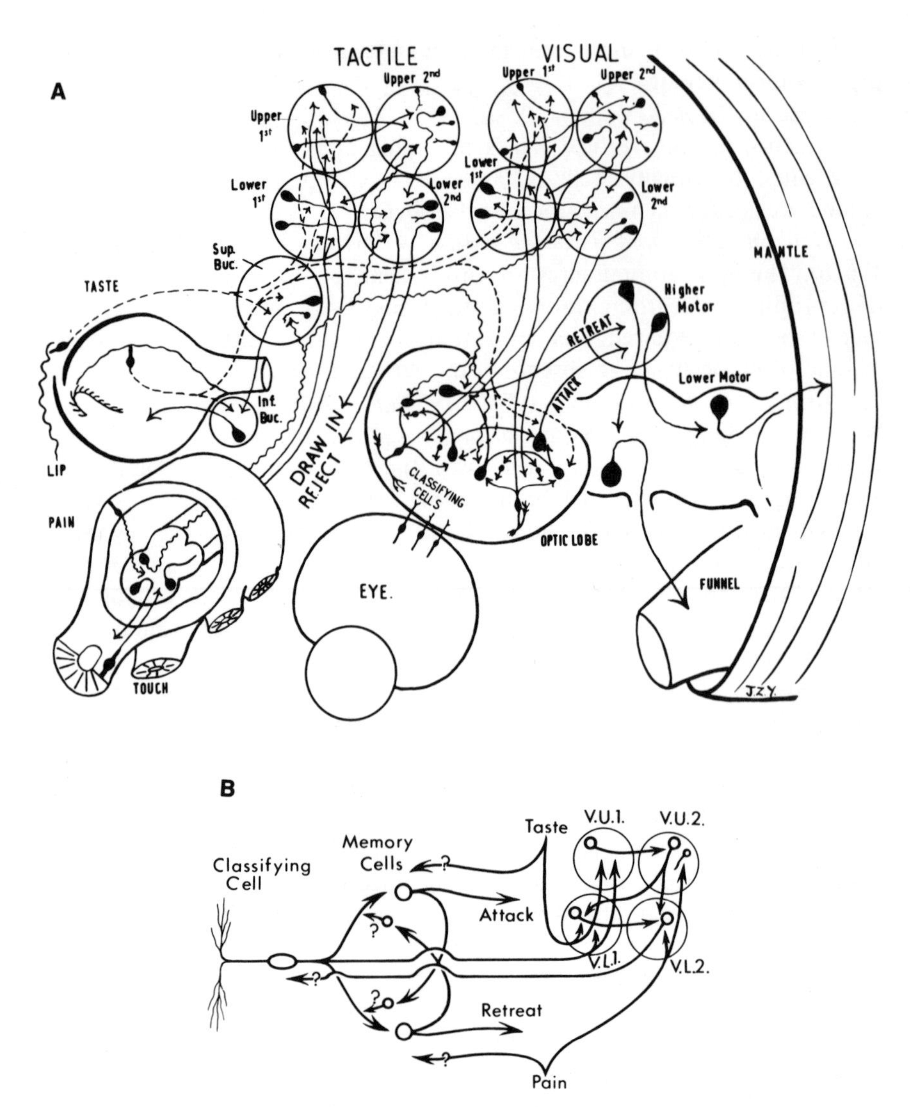

Fig. 25. A, A diagrammatic representation of the two sets of paired centers that are involved in the visual and tactile learning systems of the octopus. The pain and taste pathways have yet to be demonstrated. Tactile lobes: upper first, median inferior frontal; upper second, lateral inferior frontal; lower first, subfrontal; lower second, posterior buccal. Visual lobes: upper first, median superior frontal; upper second, vertical; lower first, lateral superior frontal; lower second, subvertical. (From Young, 1965*c*.) *Inf. Buc.*, inferior buccal lobe; *Sup. Buc.*, superior buccal lobe. B, A plan of the connections involved in "maintaining the address" of a mnemon in the optic lobe. The question marks show where evidence of the connections is lacking. (From Young, 1965*c*.) *V.U.1.*, median superior frontal lobe; *V.U.2.* vertical lobe; *V.L.1*, lateral superior frontal lobe, *V.L.2.*, subvertical lobe.

the upper visual pair, formed by the median superior frontal and vertical lobes, is seen as that of balancing the effects of food and pain in what Young (1965*c*) has called an "unless" system. The first lobe amplifies signals of positive reinforcement to promote attack unless negated by the amplification of signals of negative reinforcement in the second lobe. Most of the tracts and circuits required by this model have been identified, but the "pain" and taste inputs to the vertical lobe have yet to be demonstrated.

The vertical lobe system is also seen as providing a reexciting circuit that serves as a short-term memory to "maintain the addresses" of the classifying cells that have analyzed a stimulus until the outcome of the response can be assessed in terms of "pleasure" (food) or "pain." A plan of the connections involved in this system is given in Fig. 25B, which also illustrates the principle of Young's hypothetical memory unit, the mnemon. In this model a group of classifying cells records the occurrence of particular stimuli. Each of these cells has two outputs commanding two alternative responses, e.g., attack or retreat. Initially the system will be biased toward one response, which, in the octopus, is usually attack. The result of this response, in terms of taste or pain, is returned to the mnemon. If the outcome of the attack was food then collaterals of the "command attack" cell activate the small inhibitory neurons to block the "command retreat" pathway. Likewise, if the outcome was painful, the "command attack" pathway is inhibited. Thus a specific response will become associated with a particular stimulus.

On the basis of Young's proposals Clymer (1973) has developed a quantitative model of visual learning in the octopus for use in a computer simulation program. Some of the results generated by this model compare favorably with those obtained from experiments with octopuses. Typical learning curves and the effects of lesions to the vertical lobe were reproduced, but the model could not successfully handle reversal learning data.

Many of the results of vertical lobe removal could be understood if the lobe formed part of a short-term memory system that also served to read into and out of a long-term memory store. Disruption of the short-term memory function could explain (1) the reduced period of effectiveness of individual reinforcements; (2) the larger acquisition deficits seen with delayed reinforcement; (3) the poorer delayed-response performance; and (4) the improved acquisition with shortened intertrial intervals, all of which are seen after vertical lobe removal. Impairment of short-term memory and the readin function could account for the general acquisition deficit. Failure of postoperational performance to match that obtained prior to vertical lobe removal could result from damage to the readout function. There is then no need to postulate that some memory storage occurs in the vertical lobe itself, and much of the evidence for an inhibitor effect could be equally well explained by an acquisition deficit.

In the absence of the vertical lobe, a less effective memory could be available for the control of behavior. A defective memory could be expected to function better (1) with successive presentations of the stimuli when the attack level is not very high, e.g., when crabs are excluded from the stimulus complex and during extinction trials, and (2) with simultaneous presentation when a slightly preferred shape may be attacked at each trial, and fluctuations in the level of attack do not affect discrimination performance. These three phenomena have been demonstrated in octopuses without vertical lobes and the postulation of motivational difficulties is unnecessary to explain them. Finally, a reexciting short-term memory system could act as an amplifier by "representation," although the neuroanatomy of the vertical lobe system suggests that it could amplify signals by increasing the numbers of channels that they occupy.

VIII. ELECTROPHYSIOLOGICAL RECORDING FROM THE CEPHALOPOD NERVOUS SYSTEM

Although the first electroretinogram was recorded in *Eledone* (Beck, 1899), there have been few electrophysiological studies of cephalopods other than the work on the giant axons of squids. Of the few publications that have appeared in recent years, the majority have been concerned with the statocyst or the visual system. The practical problems encountered with these animals have undoubtedly contributed to the paucity of studies in this field. First, with the skeleton reduced to remnants of the molluscan shell and a fragile cartilaginous cranium, restraining the animal and stabilizing the electrodes present considerable difficulties. In addition, the mobile arms and the musculature of the skin tend to dislodge the electrodes, and no known curarelike agent is available. Second, the loss of blood at wounds is controlled by local vasoconstriction that may close down the supply to the area under observation. Third, the neurons are predominantly unipolar and their fibers are of small diameter. Unfortunately, it is the supraesophageal lobes, which behavioral studies have shown to be intimately involved in learning and memory, that are characterized by the presence of particularly small cells. Whereas 30% of the neurons in the lower motor centers of the subesophageal lobes have nuclear diameters of 10–20 μm, more than 99% of those in the vertical and subfrontal lobes have diameters of less than 5 μm (Young, 1963*b*). It is estimated that the vertical lobe contains some 25 million of these small amacrine cells with axons that do not leave the lobe. Output from the vertical lobe is carried by about 65,000 larger cells with nuclear diamters of 5–10 μm. In the median superior frontal lobe, from which the vertical lobe receives most of its input, half of the neuronal nuclei are less than 5 μm in diameter, while the other half fall into the 5–10 μm range.

Most investigators have tackled these problems by making extracellular recordings in acute preparations. Maturana and Sperling (1963) perfused the isolated octopus statocyst with aerated seawater and reported successful recording from single units in the middle cristal nerve. Recently, Budelmann and Wolff (1973) have made more extensive studies of the isolated statocyst. Rowell (1966) overcame the problem of restraint and blood supply by entirely removing the supraesophageal lobes from his octopuses before recording from neurons in the arms. This operation produces a placid animal that will survive several weeks despite its inability to feed. An ingenious technique that cannot be applied to studies of the supraesophageal lobes themselves! Squid and cuttlefish are more easily restrained than octopuses because they have relatively shorter arms and a large internal shell that gives some measure of rigidity to the mantle. MacNichol and Love (1960, 1961) obtained successful recordings from the retinal nerves and optic ganglion (lobe) of living *Loligo pealii* by restraining the animals in a lead jacket. When dissection was used to uncover the nerves, death occurred within a few minutes, but if the electrodes were passed through holes punched by a hypodermic needle, the squids survived indefinitely.

To my knowledge only two studies have attempted to record from the central supraesophageal lobes that are involved in the processes of learning and memory: both used *Octopus vulgaris*. Boycott *et al.* (1965) threaded insulated 100-μ copper wire electrodes through the dorsolateral part of the cranium. The octopuses survived for at least several days with about 1.5 m of these light wires trailing behind them as they moved freely in a small tank. Consistent hour-long records were obtained from the visual system of the response to flashes of light. Although many sites within the vertical and median superior frontal lobes were silent, on two occasions just discernible 20-μV waves were recorded for a brief period from in, or near, the vertical lobe. These waves appeared about 100 msec after the flash of light.

In a recent preliminary communication, Stephen (1974) reports successful extracellular recording of activity in the optic, peduncle, vertical, and median superior frontal lobes of octopuses immobilized with cold (8°C) seawater and by stitching to a solid frame. Records from the vertical lobe showed an "on" and an "off" response, which occurred 0.5 sec after the beginning and end of a 1-sec flash of light. Two further responses, occurring 1.7 sec after the original "on" and "off" responses, were also observed. When recordings were made from the median superior frontal lobe with the median superior frontal-vertical lobe tract cut, only the first "on" and "off" responses were seen. This is the first electrophysiological indication that impulses may reverberate around the vertical lobe circuit (see Fig. 25) from the median superior frontal to the vertical, subvertical, lateral superior frontal and back to the median superior frontal lobe. In view of the suggestions proposed as a result of behavioral experiments, that the vertical lobe is involved in some form of short-term memory system (see section VII, I), early confirmation of this finding is desirable.

IX. CONCLUSIONS

Learning phenomena have undoubtedly been more extensively studied in cephalopods, particularly octopuses, than in any other invertebrates. As discriminated responding can readily be established and reversed by the appropriate control of reinforcement, investigators have predominantly asked "What can an octopus learn?" or "How does the octopus learn?" rather than "Can the octopus learn?" Perhaps as a result, there have been few studies of simple classical or operant conditioning. The use of octopuses to investigate learning phenomena such as generalization, cue additivity, progressive improvement in serial reversal learning, and the overtraining reversal effect, phenomena that were originally discovered in vertebrates, has aided the development of a general theory of animal discrimination learning. The functional organization of much of the octopus nervous system has been determined and separate locations for visual and tactile learning have been identified. Given the ability of octopuses to master a variety of learning and memory tasks and the detailed descriptions of their neuroanatomy that exist, these animals are attractive subjects for studies that seek to relate adaptive behavior to neural organization.

Finally, I should like to furnish this review with an illustration of the quality of adaptive behavioral change of which octopuses are capable. Recently, I had occasion to leave a number of blind octopuses for a month, during which they were fed crabs by one of the technical staff of the Stazione Zoologica di Napoli. When feeding these octopuses on my return, I noted that many exhibited the following behavior. As the lid of their tank was opened, they would swim to the surface at the front of the tank, turn onto their backs, and spread the arms and interbrachial web like an upturned umbrella. If the crabs were dropped into the front of the tank, as the assistant had done regularly during my absence, the majority would be easily caught by the octopus. When I dropped crabs beyond the reach of the arms to the bottom of the tank, the octopus adopted a different approach. After waiting unsuccessfully in the upside-down position, it would swim to the bottom at one end of the tank and then, spreading its arms from one side of the tank to the other, move along the length of the tank collecting the crabs like a snowplow. Is there a more efficient method for a blind octopus to capture its meal?

ACKNOWLEDGMENT

It is a pleasure to thank Dr. Marion Nixon and Professor J. Z. Young, who read an earlier draft of this chapter, for their advice.

REFERENCES

Altman, J. S., and Nixon, M., 1970, Use of beaks and radula by *Octopus vulgaris* in feeding, *J. Zool. Proc. Zool. Soc. Lond., 161,* 25–38.

Arnold, J. M., and Arnold, K. O., 1969, Some aspects of hole-boring predation by *Octopus vulgaris, Am. Zool., 9,* 991–996.

Barlow, J. J., and Sanders, G. D., 1974, Intertrial interval and passive avoidance learning in *Octopus vulgaris, Anim. Learn. Behav., 2,* 86–88.

Beck, A., 1899, Uber die bei Belichtung der Netzhaut von *Eledone moschata* enstehende Actionstrome, *Arch. Ges. Physiol., 78,* 129–162.

Bierens de Haan, J. A., 1949, "Animal Psychology," Hutchinson, London, p. 119.

Bitterman, M. E., 1968, Reversal learning and forgetting, *Science* Wash., D.C., *160,* 100.

Boletzky, S. V., and Boletzky, M. V., 1969, First results in rearing *Octopus joubini* Robson 1929, *Verhandl. Naturf. Ges. Basel, 80,* 56–61.

Boycott, B. B., 1954, Learning in *Octopus vulgaris* and other cephalopods, *Pubbl. Staz. Zool. Napoli, 25,* 67–93.

Boycott, B. B., 1960, The functioning of the statocyst of *Octopus vulgaris, Proc. Soc. Lond. B. Biol. Sci., 152,* 78–87.

Boycott, B. B., 1961, The functional organisation of the brain of the cuttlefish *Sepia officinalis, Proc. R. Soc. Lond. B. Biol. Sci., 153,* 503–534.

Boycott, B. B., Lettvin, J. Y., Maturana, H. R., and Wall, P. D., 1965, Octopus optic responses, *Exp. Neurol., 12,* 247–256.

Boycott, B. B., and Young, J. Z., 1950, The comparative study of learning, *Symp. Soc. Exp. Biol., 4,* 432–453.

Boycott, B. B., and Young, J. Z., 1955*a*, Memories controlling attacks on food objects by *Octopus vulgaris* Lamarck, *Pubbl. Staz. Zool. Napoli, 27,* 232–249.

Boycott, B. B., and Young, J. Z., 1955*b*, A memory system in *Octopus vulgaris* Lamarck, *Proc. R. Soc. Lond. B. Biol. Sci., 143,* 449–480.

Boycott, B. B., and Young, J. Z., 1956, Reactions to shape in *Octopus vulgaris* Lamarck, *Proc. Zool. Soc. Lond., 126,* 491–547.

Boycott, B. B., and Young, J. Z., 1957, Effect of interference with the vertical lobe on visual discriminations in *Octopus vulgaris* Lamarck, *Proc. R. Soc. Lond. B. Sci., 146,* 439–459.

Boycott, B. B., and Young, J. Z., 1958, Reversal of learned responses in *Octopus vulgaris* Lamarck, *Anim. Behav., 6,* 45–52.

Bradley, E. A., 1974, Some observations of *Octopus joubini* Robson reared in an inlaid aquarium, *J. Zool. Proc. Zool. Soc. Lond., 173,* 355–368.

Budelmann, B.-U., and Wolff, H. G., 1973, Gravity response from angular acceleration receptors in *Octopus vulgaris*, *J. Comp. Physiol.*, *85*, 283–290.

Budelmann, B.-U., Barber, V. C., and West, S., 1973, Scanning electron microscopial studies of the arrangements and numbers of hair cells in the statocysts of *Octopus vulgaris*, *Sepia officinalis* and *Loligo vulgaris*, *Brain Res*, *56*, 25–41.

Buytendijk, F. J. J., 1933, Das Verhalten von Octopus nach teilweiser Zerstorung des "Gehirns," *Arch. Néerl. Physiol.*, *18*, 24.

Clymer, J. C., 1973, A computer simulation model of attack learning behaviour in the octopus, Technical Report no. 141, Computer and Communication Sciences Department, University of Michigan.

Cousteau, J. Y., and Diole, P., 1973, "Octopus and Squid: Soft Intelligence," Cassel and Co. Ltd., London.

Crancher, P., King, M. G., Bennett, A., and Montgomery, R. B., 1972, Conditioning of a free operant in *Octopus cyaneus* Gray, *J. Exp. Anal. Behav.*, *17*, 359–362.

Denton, E. J., 1974, On the buoyancy and the lives of modern and fossil cephalopods, *Proc. R. Soc. Lond. B. Biol. Sci.*, *185*, 273–299.

Denton, E. J., and Gilpin-Brown, J. B., 1973, Floatation mechanisms in modern and fossil cephalopods, *in* "Advances in Marine Biology" (F. S. Russell and M. Yonge, eds.), Vol. 11, pp 197–268, Academic Press, New York.

Deutsch, J. A., 1955, A theory of shape recognition, *Br. J. Psychol.*, *46*, 30–37.

Deutsch, J. A., 1960*a*, The plexiform zone and shape recognition in *Octopus*, *Nature (Lond.)*, *185*, 443–446.

Deutsch, J. A., 1960*b*, Theories of shape discrimination in *Octopus*, *Nature* (*Lond.*) *188*, 1090–1092.

Dews, P. M., 1959, Some observations on an operant in the octopus, *J. Exp. Anal. Behav.*, *2*, 57–63.

Dijkgraaf, S., 1963, Versuche, über Schallwahrnehmung bei Tintenfischen, *Naturwissenschaften*, 50, 50.

Dilly, P. N., 1963, Delayed responses in *Octopus*, *J. Exp. Biol.*, *40*, 393–401.

Dodwell, P. C., 1961, Facts and theories of shape discrimination, *Nature* (Lond.), *191*, 578–580.

Eninger, M. U., 1952, Habit summation in a selective learning problem, J. Comp. Physiol. Psychol., 45, 604–608.

Goldsmith, M., 1917*a*, Quelques réactions sensorielles chez le Poulpe, *C. R. Acad. Sci. Paris*, *164*, 448.

Goldsmith, M., 1917*b*, Acquisition d'une habitude chez le Poulpe, *C. R. Acad. Sci. Paris*, *164*, 737.

Goldsmith, M., 1917*c*, Quelques réactions du Poulpe; contribution à la psychologie des invertébrés, *Bull. Inst. Gén. Psychol.*, *17*, 25–44.

Gonzales, R. C., Behrend, E. R., and Bitterman, M. E., 1967, Reversal learning and forgetting in bird and fish, *Science*, *158*, 519–521.

Graziadei, P., 1964, Electron microscopy of some primary receptors in the sucker of *Octopus vulgaris*, *Z. Zellforsch. Mikrosk. Anat. Abt. Histochem.*, *64*, 510–522.

Graziadei, P., 1965, Sensory receptor cells and related neurons in cephalopods, *Cold Spring Harbor Symp. Quant. Biol.*, *30*, 45–57.

Graziadei, P., 1971, The nervous system of the arms, Ch. 3. *in* "The Anatomy of the Nervous System of *Octopus vulgaris*" (Young, J. Z.), Clarendon Press, Oxford.

Grice, G. R., 1948, The relation of secondary reinforcement to delayed reward in visual discrimination learning, *J. Exp. Psychol.*, *38*, 1–16.

Hubbard, S. J., 1960, Hearing and the octopus statocyst, *J. Exp. Biol.*, *37* 845–853.

Hubel, D. H., and Wiesel, T. N., 1962, Receptive fields, binocular interaction and functional architecture in the cat's visual cortex, *J. Physiol.* (Lond.), *160*, 106–154.

Itami, K., Izawa, Y., Maeda, S., and Nakai, K., 1963, Notes on the laboratory culture of the octopus larvae, *Bull. Jap. Soc. Sci. Fish., 29*, 514–520.

Kovačević, N., and Rakić, L. J., 1971, Circadian rhythm and visual discrimination in *Octopus vulgaris* Lamarck, *Arh. Biol. Nauka, 23*, 3–4.

Kühn, A., 1930, Über Farbensinn and Anpassung der Körpefarbe an die Umgebung bei Tintenfischen, *Nachr. Ges. Wiss. Gottingen, Mathnat. Kl.*, 10.

Kühn, A., 1950, Über Farbwechsel und Farbensinn von Cephalopoden. *Z. Vgl. Physiol., 32*, 572.

Lane, F. W., 1957, "Kingdom of the Octopus," Jarrolds Publishers Ltd., London.

Lawrence, D. H., 1952, The transfer of a discrimination along a continuum, *J. Comp. Physiol. Psychol., 45*, 511–516.

Mackintosh, J., 1962, An investigation of reversal learning in *Octopus vulgaris* Lamarck, *Q. J. Exp. Psychol., 14*, 15–22.

Machintosh, N. J., 1965*a*, Discrimination learning in the octopus, *Anim. Behav.* (Suppl. No. 1), 129–134.

Mackintosh, N. J., 1965*b*, Selective attention in animal discrimination learning, *Psychol. Bull., 64*, 124–150.

Mackintosh, N. J., and Mackintosh, J., 1963, Reversal learning in *Octopus vulgaris* Lamarck with and without irrelevant cues, *Q. J. Exp. Psychol., 15*, 236–242.

Mackintosh, N. J., and Mackintosh, J., 1964*a*, Performance of *Octopus* over a series of reversals of a simultaneous discrimination, *Anim. Behav., 12*, 321–324.

Mackintosh, N. J., and Mackintosh, J., 1964*b*, The effect of overtraining on a non-reversal shift in *Octopus*, *J. Genet. Psychol., 106*, 373–377.

MacNichol, E. F., and Love, W. E., 1960, Electrical response of the retinal nerve and optic ganglion of the squid, *Science, 132*, 737–738.

MacNichol, E. F., Jr., and Love, W. E., 1961, Impulse discharges from the retinal nerve and optic ganglion of the squid, *in* "Symposium of the Visual System" R. Jung and H. Kornhuber, eds., pp. 97–103, Springer-Verlag, Berlin.

Maldonado, H., 1963*a*, The positive learning process in *Octopus vulgaris*, *Z. Vgl. Physiol., 47*, 191–214.

Maldonado, H., 1963*b*, The general amplification function of the vertical lobe in *Octopus vulgaris*, *Z. Vgl. Physiol., 47*, 215–229.

Maldonado, H., 1963*c*, The visual attack learning system in *Octopus vulgaris*, *J. Theor. Biol., 5*, 470–488.

Maldonado, H., 1964, The control of attack by *Octopus*, *Z. Vgl. Physiol., 47*, 656–674.

Maldonado, H., 1965, The positive and negative learning process in *Octopus vulgaris* Lamarck: Influence of the vertical and median superior frontal lobes, *Z. Vgl. Physiol., 51*, 185–203.

Maldonado, H., 1968, Effect of electroconvulsive shock on memory in *Octopus vulgaris* Lamarck, *Z. Vgl. Physiol., 59*, 25–37.

Maldonado, H., 1969, Further investigations on the effect of electroconvulsive shock (ECS) on memory in *Octopus vulgaris*, *Z. Vgl. Physiol., 63*, 113–118.

Mangold, K., and von Boletzky, S., 1973, New data on reproductive biology and growth of *Octopus vulgaris*, *Marine Biol. N.Y., 19*, 7–17.

Mangold-Wirz, K., 1969, The swimming of pelagic squids, *Documenta Geigy Nautilus, 7*, 6–7.

Maturana, H. R., and Sperling, S., 1963, Unidirectional response to angular acceleration recorded from the middle cristal nerve in the statocyst of *Octopus vulgaris*, *Nature* (Lond.), *197*, 815–816.

Mello, N. K., 1965*a*, Interhemispheric reversal of mirror-image oblique lines after monocular training in pigeons, *Science, 148*, 252–254.

Mello, N. K., 1965*b*, Mirror-image reversal in pigeons, *Science* Wash., D.C., *149*, 1519–1520.

Messenger, J. B., 1968, The visual attack of the cuttlefish, *Sepia officinalis, Anim. Behav., 16*, 342–357.

Messenger, J. B., 1971, Two-stage recovery of a response in *Sepia, Nature* (Lond.), *232*, 202–203.

Messenger, J. B., 1973*a*, Learning in the cuttlefish, *Sepia, Anim. Behav., 21*, 801–826.

Messenger, J. B., 1973*b*, Learning performance and brain structure: A study in development, *Brain Res. 58*, 519–523.

Messenger, J. B., 1974, Reflecting elements in the skin of cephalopods and their importance for camouflage, *J. Zool.* (Lond.), *174*, 387–395.

Messenger, J. B., and Sanders, G. D., 1971, The inability of *Octopus vulgaris* to discriminate monocularly between oblique rectangles, *Int. J. Neurosci., 1*, 171–173.

Messenger, J. B., and Sanders, G. D., 1972, Visual preference and two cue discrimination learning in octopus, *Anim. Behav., 20*, 580–585.

Messenger, J. B., Wilson, A. P., and Hedge, A., 1973, Some evidence for colour blindness in *Octopus, J. Exp. Biol., 59*, 77–94.

Mikhailoff, S., 1920, Expériences reflexologiques; expérience nouvelles sur *Eledone moschata, Bull. Inst. Oceanogr.* Monaco, *379*, 1–11.

Moody, M. F., 1962, Evidence for the intraocular discrimination of vertically and horizontally polarised light by *Octopus, J. Exp. Biol., 39*, 21–30.

Moody, M. F., and Parriss, J. R., 1961, The discrimination of polarized light by *Octopus*, a behavioural and morphological study, *Z. Vgl. Physiol., 44*, 268–291.

Muntz, W. R. A., 1961*a*, Interocular transfer in *Octopus vulgaris, J. Comp. Physiol. Psychol., 54*, 49–55.

Muntz, W. R. A., 1961*b*, The function of the vertical lobe system of *Octopus* in interocular transfer, *J. Comp. Physiol. Psychol., 54*, 186–191.

Muntz, W. R. A., 1961*c*, Interocular transfer in *Octopus*: Bilaterality of the engram, *J. Comp. Physiol. Psychol., 54*, 192–195.

Muntz, W. R. A., 1962, Stimulus generalization following monocular training in *Octopus, J. Comp. Physiol. Psychol., 55*, 535–540.

Muntz, W. R. A., 1963, Intraretinal transfer and the function of the optic lobes in *Octopus, Q. J. Exp. Psychol., 15*, 116–124.

Muntz, W. R. A., 1970, An experiment on shape discrimination and signal detection in octopus, *Q. J. Exp. Psychol., 22*, 82–90.

Muntz, W. R. A., Sutherland, N.S., and Young, J. Z., 1962, Simultaneous shape discrimination in *Octopus* after removal of the vertical lobe, *J. Exp. Biol., 39*, 557–566.

Nixon, M., 1969, The lifespan of *Octopus vulgaris* Lamarck, *Proc. Malacol. Soc. Lond., 38*, 529–540.

Nixon, M., and Young, J. Z., 1966, Levels of responsiveness to food or its absence and the vertical lobe circuit of *Octopus vulgaris* Lamarck, *Z. Vgl. Physiol., 53*, 165–184.

Noble, J., 1966, Mirror-images and the forebrain commissures of the monkey, *Nature* (Lond.), *211*, 1263–1265.

Ohshima, Y., and Sang Choe, 1961, On the rearing of young cuttlefish and squid, *Bull. Jap. Soc. Sci. Fish., 27*, 979–986.

Ohshima, Y., and Sang Choe, 1963, Rearing of cuttlefish and squids, *Nature* (Lond.), *197, 307.*

Packard, A., 1961, Sucker display of Octopus, Nature (Lond.), *190*, 736–737.

Packard, A., 1969, Visual acuity and eye growth in *Octopus vulgaris* (Lamarck), *Monit. Zool. Ital., 3*, 19–32.

Packard, A., 1972, Cephalopods and fish: The limits of convergence, *Biol. Rev.* (Camb.), *47*, 241–307.

Packard, A., and Albergoni, V., 1970, Relative growth, nucleic acid content and cell numbers of the brain in *Octopus vulgaris* (Lamarck), *J. Exp. Biol., 52,* 539–553.

Packard, A., and Sanders, G. D., 1969, What the octopus shows to the world, *Endeavour* Engl. Ed., *28,* 92–99.

Packard, A., and Sanders, G. D., 1971, Body patterns of *Octopus vulgaris* and maturation of the response to disturbance, *Anim. Behav., 19,* 780–790.

Parriss, J. R., 1963, Interference in learning and lesions in the visual system of *Octopus vulgaris, Behaviour, 21,* 233–245.

Pieron, H., 1911, Contribution à la physchologie du poulpe; l'acquisition d'habitudes, *Bull. Inst. Gén. Psychol., 17,* 111–119.

Pieron, H., 1914, Contribution à la psychologie du poulpe; la mémoire sensorielle, *L'Année Psychol., 20,* 182.

Pilson, M. E. Q., and Taylor, P. B., 1961, Hole drilling by *Octopus, Science, 134,* 1366–1368.

Polimanti, C., 1910, Les cephalopodes ont-ils une mémoire? *Arch. Psychol., 10,* 84–87.

Prescott, J. H., and Brosseau, C., 1962, Transportation and display of the giant pacific octopus, *Octopus appollyon, in International Zoology Year Book, 6,* Jarvis and Morris, eds., 53–57.

Rhodes, J. M., 1963, Simultaneous discrimination in Octopus, *Pubbl. Staz. Zool. Napoli, 33,* 83–91.

Rowell, C. H. F., 1966, Activity of interneurones in the arm of *Octopus* in response to tactile stimulation, *J. Exp. Biol., 44,* 589–605.

Rowell, C. H. F., and Wells, M. J., 1961, Retinal orientation and the discrimination of polarised light by octopuses, *J. Exp. Biol., 38,* 827–831.

Sanders, G. D., 1970*a*, The retention of visual and tactile discriminations by *Octopus vulgaris*, Ph.D. thesis, University of London.

Sanders, G. D., 1970*b*, Long-term memory of a tactile discrimination in *Octopus vulgaris* and the effect of vertical lobe removal, *Brain Res., 20,* 59–73.

Sanders, G. D., 1970*c*, Long-term tactile memory in *Octopus*: Further experiments on the effect of vertical lobe removal, *Brain Res., 24,* 169–178.

Sanders, G. D., and Barlow, J. J., 1971, Variations in retention performance during long-term memory formation, *Nature* (Lond.), *232,* 203–204.

Sanders, F. K., and Young, J. Z., 1940, Learning and other functions of the higher nervous centre of *Sepia, J. Neurophysiol., 3,* 501–526.

Schiller, P. H., 1948, Studies on learning in the octopus, *Report of Committee on Research for the National Academy of Sciences,* 158–160.

Schiller, P. H., 1949, Delayed detour response in the octopus, *J. Comp. Physiol Psychol., 42,* 220–225.

Stephen, R. O., 1974, Electrophysiological studies of the brain of *Octopus vulgaris, J. Physiol. 240,* 19P–20P.

Sutherland, N. S., 1957*a*, Visual discrimination of orientation by octopus, *Brit. J. Psychol., 48,* 55–71.

Sutherland, N. S., 1957*b*, Visual discrimination of orientation and shape by the octopus, *Nature* Lond., *179,* 11–13.

Sutherland, N. S., 1958*a*, Visual discrimination of shape by octopus. Squares and triangles, *Q. J. Exp. Psychol., 10,* 40–47.

Sutherland, N. S., 1958*b*, Visual discrimination of the orientation of rectangles by *Octopus vulgaris* Lamarck, *J. Comp. Physiol. Psychol., 51,* 452–458.

Sutherland, N. S., 1959*a*, Visual discrimination of shape by octopus. Circles and squares, and circles and triangles, *Q. J. Exp. Psychol., 11,* 24–32.

Sutherland, N. S., 1959*b*, A test of a theory of shape discrimination in *Octopus vulgaris* Lamarck, *J. Comp. Physiol. Psychol.*, *52*, 135–141.

Sutherland, N. S., 1959*c*, Stimulus analysing mechanisms, *in* "Proceeding of a symposium on the mechanisation of Thought Processes," Vol. 2, H.M. Stationery Office, London.

Sutherland, N. S., 1960*a*, Visual discrimination of orientation by octopus: Mirror images, *Brit. J. Psychol.*, *51*, 9–18.

Sutherland, N. S., 1960*b*, The visual discrimination of shape by octopus: Squares and rectangles, *J. Comp. Physiol. Psychol.*, *53*, 95–103.

Sutherland, N. S., 1960*c*, Visual discrimination of shape by octopus: Open and closed forms, *J. Comp. Physiol. Psychol.*, *53*, 104–112.

Sutherland, N. S., 1960*d*, The visual system of *Octopus* (3) Theories of shape discrimination in *Octopus*, *Nature Lond.*, *186*, 840–844.

Sutherland, N. S., 1960*e*, Theories of shape discrimination in *Octopus*, *Nature* (Lond.), *188*, 1092–1094.

Sutherland, N. S., 1961, Discrimination of horizontal and vertical extents by *Octopus*, *J. Comp. Physiol. Psychol.*, *54*, 43–48.

Sutherland, N. S., 1963*a*, Visual acuity and discrimination of stripe widths in *Octopus vulgaris* Lamarck, *Pubbl. Staz. Zool. Napoli*, *33*, 92–109.

Sutherland, N. S., 1963*b*, Shape discrimination and receptive fields, *Nature* (Lond.), *197*, 118–122.

Sutherland, N. S., 1968, Outlines of a theory of visual pattern recognition in animals and men, *Proc. R. Soc. Lond. B Biol. Sci.*, *171*, 297–317.

Sutherland, N. S., 1969, Shape discrimination in rat, octopus and goldfish: A comparative study, *J. Comp. Physiol. Psychol.*, *67*, 160–176.

Sutherland, N. S., and Holgate, V., 1966, Two-cue discrimination learning in rats. *J. Comp. Physiol. Psychol.*, *61*, 198–207.

Sutherland, N. S., and Mackintosh, N. J., 1971, "Mechanisms of Animal Discrimination Learning," Academic Press, New York and London.

Sutherland, N. S., Mackintosh, N. J., and Mackintosh, J., 1963*a*, Simultaneous discrimination training of *Octopus* and transfer of discrimination along a continuum, *J. Comp. Physiol. Psychol.*, *56*, 150–156.

Sutherland, N. S., Mackintosh, J., and Mackintosh, N. J., 1963*b*, The visual discrimination of reduplicate patterns by octopus, *Anim. Behav.*, *11*, 106–110.

Sutherland, N. S., Mackintosh, N. J., and Mackintosh, J., 1965, Shape and size discrimination in *Octopus:* The effects of pretraining along different dimensions, *J. Gen. Psychol.*, *106*, 1–10.

Sutherland, N. S., and Muntz, W. R. A., 1959, Simultaneous discrimination training and preferred directions of motion in visual discrimination of shape in *Octopus vulgaris* Lamarck, *Pubbl. Staz. Zool. Napoli*, *31*, 109–126.

Ten Cate, J., and Ten Cate-Kazeewa, B., 1938, Les *Octopus vulgaris* peuvent-ils discerner les formes? *Arch. Néerl. Physiol.*, *23*, 541–551.

Thomas, R. F., & Opresko, L., 1973, Observations on *Octopus joubini:* Four laboratory reared generations, *The Nautilus*, *87*, 61–65.

von Uexkull, J., 1905, "Leitfaden in das Studium der exp. Biologie der Wassertiere," p. 151, *Wiesbaden.* (Quoted by Polimanti, C., 1910.)

Walker, J. J., Longo, N., & Bitterman, M. E., 1970, The octopus in the laboratory. Handling, maintenance and training, *Behav. Res. Methods Instrum.*, *2*, 15–18.

Wells, M. J., 1959*a*, A touch learning centre in Octopus, *J. Exp. Biol.*, *36*, 590–612.

Wells, M. J., 1959*b*, Functional evidence for neurone fields representing the individual arms within the central nervous system of *Octopus*, *J. Exp. Biol.*, *36*, 501–511.

Wells, M. J., 1960, Proprioception and visual discrimination of orientation in *Octopus, J. Exp. Biol., 37,* 489–499.
Wells, M. J., 1961*a*, Weight discrimination by *Octopus, J. Exp. Biol., 38,* 127–133.
Wells, M. J., 1961*b*, Centres for tactile and visual learning in the brain of *Octopus, J. Exp. Biol., 38,* 811–826.
Wells, M. J., 1962*a*, "Brain and Behaviour in Cephalopods," Heinemann, London.
Wells, M. J., 1962*b*, Early learning in *Sepia, Symp. Zool. Soc. London, 8,* 146–169.
Wells, M. J., 1963, Taste by touch; some experiments with *Octopus, J. Exp. Biol., 40,* 187–193.
Wells, M. J., 1964*a*, Tactile discrimination of shape by *Octopus, J. Exp. Psychol., 16,* 156–162.
Wells, M. J., 1964*b*, Tactile discrimination of surface curvature and shape by the octopus, *J. Exp. Biol., 41,* 433–445.
Wells, M. J., 1964*c*, Detour experiments with octopus, *J. Exp. Biol., 41,* 621–642.
Wells, M. J., 1965*a*, Learning in the octopus, *Symp. Soc. Exp. Biol., 20,* 477–507.
Wells, M. J., 1965*b*, Learning and movement in octopuses, *Anim. Behav.* (Suppl. No.1.), 115–128.
Wells, M. J., 1965*c*, Learning by marine invertebrates, *Adv. Mar. Biol., 3,* 1–62.
Wells, M. J., 1965*d*, The vertical lobe and touch learning in the octopus, *J. Exp. Biol., 42,* 233–255.
Wells, M. J., 1966, "The brain and Behaviour of Cephalopods. Physiology of Mollusca," K. M. Wilbur and C. M. Younge (eds.), Vol. 2, Ch.15, pp. 547–590, Academic Press, New York and London.
Wells, M. J., 1967, Short-term learning and interocular transfer in detour experiments with octopuses, *J. Exp. Biol., 47,* 383–408.
Wells, M. J., 1970, Detour experiments with split-brain octopuses, *J. Exp. Biol., 53,* 375–389.
Wells, M. J., and Wells, J., 1956, Tactile discrimination and the behavior of blind *Octopus, Pubbl. Staz. Zool. Napoli, 28,* 94–126.
Wells, M. J., and Wells, J., 1957*a*, Repeated presentation experiments and the function of the vertical lobe in *Octopus, J. Exp. Biol., 34,* 469–477.
Wells, M. J., and Wells, J., 1957*b*, The function of the brain of *Octopus* in tactile discrimination, *J. Exp. Biol., 34,* 131–142.
Wells, M. J., and Wells, J., 1957*c*, The effect of lesions to the vertical and optic lobes on tactile discrimination in *Octopus, J. Exp. Biol. 34,* 378–393.
Wells, M. J., and Wells, J., 1958*a*, The effect of vertical lobe removal on the performance of octopuses in retention tests, *J. Exp. Biol., 35,* 337–348.
Wells, M. J., and Wells, J., 1958*b*, The influence of preoperational training on the performance of octopuses following vertical lobe removal, *J. Exp. Biol., 35,* 324–336.
Wells, M. J., and Young, J. Z., 1965, Split-brain preparations and touch learning in the *Octopus, J. Exp. Biol., 43,* 565–579.
Wells, M. J., and Young, J. Z., 1966, Lateral interaction and transfer in the tactile memory of the *Octopus, J. Exp. Biol., 45,* 383–400.
Wells, M. J., and Young, J. Z., 1968*a*, Changes in textural preferences in *Octopus* after lesions, *J. Exp. Biol., 49,* 401–412.
Wells, M. J., and Young, J. Z., 1968*b*, Learning with delayed rewards in *Octopus, Z. Vgl. Physiol., 61,* 103–128.
Wells, M. J., and Young, J. Z., 1969*a*, Learning at different rates of training in the octopus, *Anim. Behav., 17,* 406–415.
Wells, M. J., and Young, J. Z., 1969*b*, The effect of splitting part of the brain or removal of the median inferior frontal lobe on touch learning in *Octopus, J. Exp. Biol., 50,* 515–526.

Wells, M. J., and Young, J. Z., 1970*a*, Stimulus generalisation in the tactile system of *Octopus*, *J. Neurobiol.*, *2*, 31–46.
Wells, M. J., and Young, J. Z., 1970*b*, Single session learning by octopuses, *J. Exp. Biol.*, *53*, 779–788.
Wells, M. J., and Young, J. Z., 1972, The median inferior frontal lobe and touch learning in the *Octopus*, *J. Exp. Biol.*, *56*, 381–402.
Wirz, K., 1954, Etudes quantitatives sur le système nerveux des Cèphalopodes. *C. R. Acad. Sci. Paris*, *238*, 1353–1355.
Wolterding, M. R., 1971, The rearing and maintenance of *Octopus briareus* in the laboratory, with aspects of their behavior and biology, M.S. thesis, University of Miami, Florida.
Young, J. Z., 1956, Visual responses by *Octopus* to crabs and other figures before and after training, *J. Exp. Biol.*, *33*, 709–729.
Young, J. Z., 1958*a*, Response of untrained octopuses to various figures and the effects of vertical lobe removal, *Proc. R. Soc. Lond. B. Biol. Sci.*, *149*, 463–483.
Young, J. Z., 1958*b*, Effect of removal of various amounts of the vertical lobes on visual discrimination by *Octopus*, *Proc. R. Soc. Lond. B. Biol. Sci.*, *149*, 441–462.
Young, J. Z., 1959, Extinction of unrewarded responses in *Octopus*, *Pubbl. Staz. Zool. Napoli*, *31*, 225–247.
Young, J. Z., 1960*a*, The statocysts of *Octopus vulgaris*, *Proc. R. Soc. Lond. B. Biol. Sci.*, *152*, 3–29.
Young, J. Z., 1960*b*, Unit processes in the formation of representations in the memory of *Octopus*, *Proc. R. Soc. Lond. B. Biol. Sci.*, *153*, 1–17.
Young, J. Z., 1960*c*, The visual system of Octopuses. (1) Regularities in the retina and optic lobes of *Octopus* in relation to form discrimination, *Nature* (Lond.), *186*, 836–839.
Young, J. Z., 1960*d*, Failures of discrimination learning following the removal of the vertical lobes in *Octopus*, *Proc. R. Soc. Lond. B. Biol. Sci.*, *153*, 18–46.
Young, J. Z., 1961*a*, Learning and discrimination in the octopus, *Biol. Rev.* (Camb.), *36*, 32–96. 32–96.
Young, J. Z., 1961*b*, Rates of establishment of representations in the memory of octopuses with and without vertical lobes, *J. Exp. Biol.*, *38*, 43–60.
Young, J. Z., 1962*a*, The optic lobes of *Octopus vulgaris*, *Philos. Trans. B.*, *245*, 19–58.
Young, J. Z., 1962*b*, Reversal of learning in *Octopus* and the effect of removal of the vertical lobe, *Q. J. Exp. Phychol.*, *14*, 193–205.
Young, J. Z., 1962*c*, Repeated reversal of training in *Octopus*, *Q. J. Exp. Psychol.*, *14*, 206–222.
Young, J. Z., 1963*a*, Light-and-dark-adaption in the eyes of some cephalopods, *Proc. Zool. Soc. Lond.*, *140*, 255–272.
Young, J. Z., 1963*b*, The number and sizes of nerve cells in *Octopus*, *Proc. Zool. Soc. London.*, *140*, 229–254.
Young, J. Z., 1963*c*, Some essentials of neural memory mechanisms. Paired centres that regulate and address the signals of the results of action, *Nature* (Lond.), *198*, 626–630.
Young, J. Z., 1964*a*, "A Model of the Brain," *Clarendon Press, Oxford.*
Young, J. Z., 1964*b*, Paired centres for the control of attack by *Octopus*, *Proc. R. Soc. Lond. B. Biol. Sci.*, *159*, 568–588.
Young, J. Z., 1965*a*, Influence of previous preferences on the memory of *Octopus vulgaris* after removal of the vertical lobe, *J. Exp. Biol.*, *43*, 595–603.
Young, J. Z., 1965*b*, Two memory stores in one brain, *Endeavour Engl. Ed.*, *24*, 13–20.
Young, J. Z., 1965*c*, The organisation of a memory system, *Proc. R. Soc. Lond. B. Biol. Sci.*, *163*, 285–320.

Young, J. Z., 1968, Reversal of a visual preference in *Octopus* after removal of the vertical lobe, *J. Exp. Biol.*, *49*, 413–419.

Young, J. Z., 1970, Short and long memories in *Octopus* and the influence of the vertical lobe system, *J. Exp. Biol.*, *52*, 385–393.

Young, J. Z., 1971, "The Anatomy of the Nervous System of *Octopus vulgaris*," Clarendon Press, Oxford.

Chapter 12

THE ECHINODERMS[1]

A. O. D. WILLOWS

Friday Harbor Laboratories, University of Washington
Seattle, Washington

AND

W. C. CORNING

Department of Psychology, University of Waterloo
Waterloo, Ontario, Canada

Superficial observation of the echinoderms might lead one to assume that these animals are not only sluggish and clumsy in their behavior, but also extremely inefficient in the execution of their life activities. Careful investigation, however, has shown that these organisms behave in a manner which is highly satisfactory for their habitat and mode of life. With their rather extraordinary equipment and their peculiar habits, speed of movement is not necessary for an effective execution of the hazardous business of living. By means of their muscular walls, their tube feet, and their pedicellariae, many groups of these creatures are able to carry out a wide range of manipulatory responses, activities which could easily be the subject of envy on the part of man. With these organs the animal keeps the integument clear of dirt and incrusting organism; also, it captures active prey, overpowers it, and transports it over any part of the body surface toward the region of the mouth. It is very much as if the entire body of a human being were bristling with hundreds of hands varying greatly in structure and complexity to meet the diverse needs of the organism (Warden *et al.*, 1940, p. 412).

I. INTRODUCTION

That the starfish, sea urchins, and their relatives are capable of a range of interesting behaviors, and perhaps even of "learning," will come as a distinct surprise to some. The phylum is a diverse but entirely marine one containing many bizarre forms, many that are more-or-less immobile and

[1] Preparation of this paper was aided in part by a grant from The National Research Council.

plantlike in this regard, and numerous genera that are very poorly known in terms of their physiology and behavior.

While most echinoderms have a characteristic radial symmetry built around five nearly identical sectors (Fig. 1), there are numerous examples of starfish (asteroids) and sea lilies (crinoids), in particular, with nonpentagonal arrangements and some with as many as 200 radii. Members of the phylum are present from the high intertidal to the abyssal depths and, in places, represent the majority of the biomass, e.g., crinoids have been observed to overlap one another and to cover the sea floor many animals deep in places.

A nearly universal feature of the pentaradial symmetry in all echinoderms is the rows of highly mobile tube feet that serve in many as the primary means of locomotion. It was the presence of tube feet and certain internal structures (particularly a perivisceral coelom) along with their unique bilaterally symmetric larval forms that led zoologists to separate the

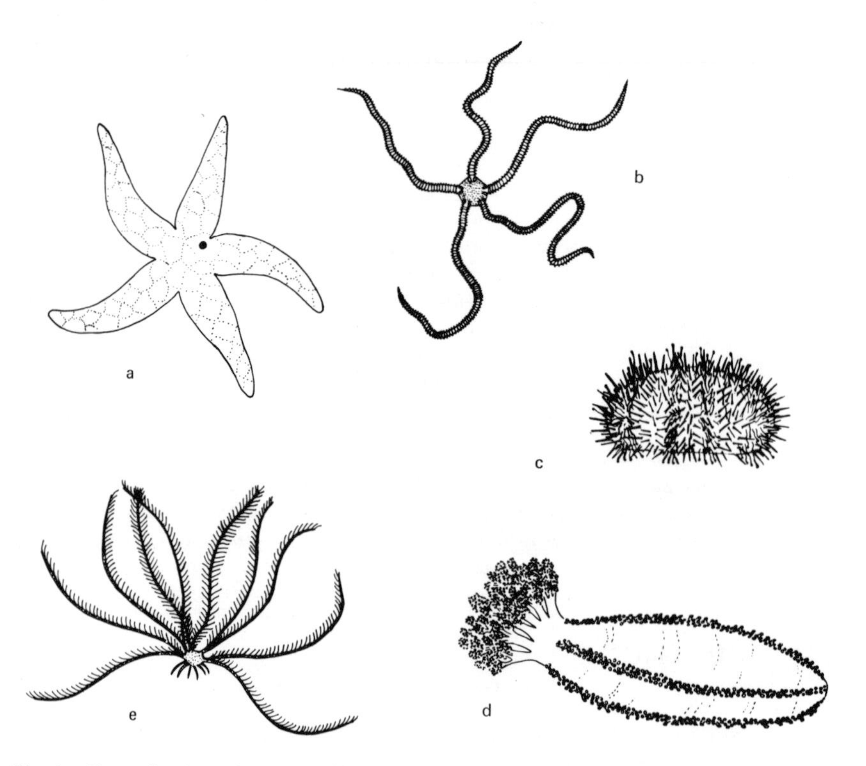

Fig. 1. Examples from the five major living classes of echinoderms: a, The starfish *Pisaster*; b, Ophiuroidea, the brittle star *Ophiura*; c, Echinoidea, the sea urchin *Strongylocentrotus*; d, Holothuroidea, the sea cucumber *Cucumaria*; e, Crinoidea, the sea lily *Antedon*.

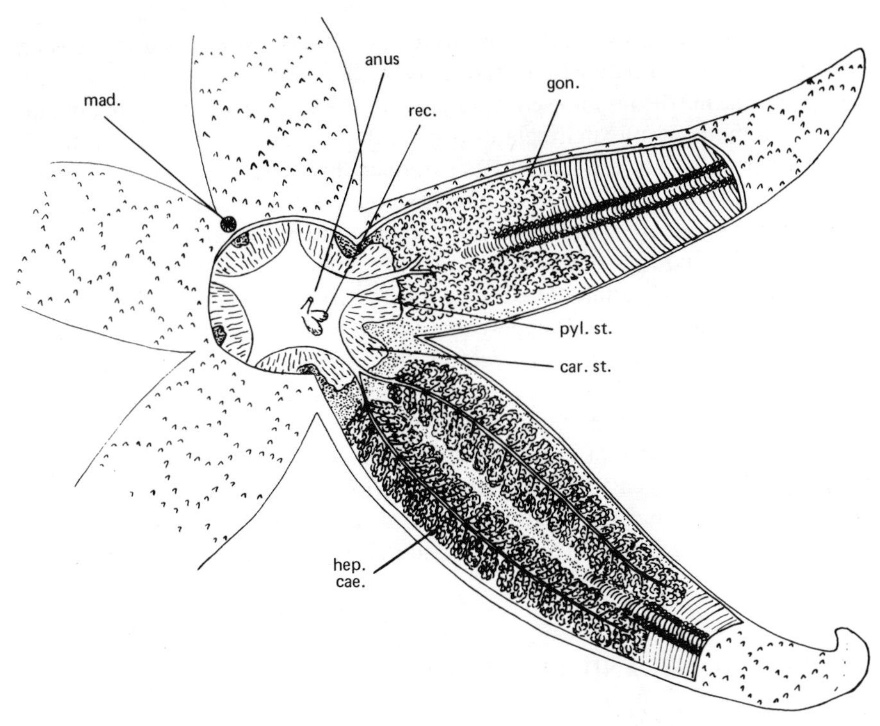

Fig. 2. The major internal anatomical structures and organ systems, excepting the water-vascular system, in the asteroids; *car st.*, cardiac stomach; *gon.*, gonad; *hep. cae.*, hepatic caeca; *mad.*, madreporite; *pyl. stom.*, pyloric stomach; *rec.*, rectal caeca.

echinoderms from the other radially symmetric group, the coelenterates. Up to that time both had been classified by Lamarck in the phylum Radiata.

The echinoderms have a distinct organ level of organization with well-defined digestive tract, vascular system, reproductive organs, nervous system, and respiratory and feeding structures (Fig. 2). Many have remarkable powers of regeneration. Fishermen trying to eradicate predatory starfish from bivalve beds have been known to gather up the starfish and tear them apart and toss them back. The end result of this procedure was that most of the pieces regenerated whole starfish and grossly multiplied the predation. Unfortunately, from the point of view of learning studies, regeneration takes place very slowly (over months or years) and is therefore not likely to be useful in experiments that one might wish to design to analyze the specificity of sites involved in control of behavior and behavioral plasticity.

Several features characterize and define the phylum. For instance, the hydrocoel (of which the tube feet are external protrusions) is an extension of the perivisceral coelom and serves the functions of locomotion, respiration, nutrition, and sensation. As well, all echinoderms have an exoskeleton made up of calcite plates (ossicles) or spicules scattered in a tough skin. Virtually all echinoderms reproduce by separate sexes, releasing enormous numbers (10^6–10^7) of gametes freely into the sea. The larvae are bilaterally symmetric and undergo a drastic alteration at metamorphosis, whereby left and right sides of the larva become oral and aboral sides in the radially symmetric adult. The larvae have a developmental sequence and a morphology that strongly suggest that they may be ancestral to the animals of the chordate line. For this reason, and because of the very significant role that the echinoderms play in most marine environments, their behavior and learning capabilities deserve review.

General reviews of echinoderm biology are widely available, and the reader is encouraged to see accounts by Nichols (1962), Clark (1962), and Hyman (1955), and the specific details of echinoderm physiology and behavior in the work of Binyon (1972) and that edited by Boolootian (1966).

II. TAXONOMY AND EVOLUTION

A. Classification

The principal schemes in use for the systematic classification of the phylum Echinodermata agree that there are ten classes in the group. Of these, five are extinct. The remaining five classes are (see Fig. 1) the following.

Crinoidea

These include the sea lilies and feather stars. The former are found exclusively at great depths (100–4500 fathoms) and usually have a rootlike base of cirri and a long stalk leading up to the main body structures and arms. Virtually nothing is known of the behavior or physiology of these forms because most dredged specimens are badly damaged and incomplete.

The sea feathers are not stalked, although they do have a base of rootlike cirri by which the animals may be attached to the substratum. Unlike the sea lilies, they are found in shallow waters. Commonly, they have five pairs of arms and each arm is lined laterally on both sides by a row of short, narrow, leaflike extensions called "pinnules." Both mouth and anus are on the uppermost side, contrary to the situation found in the other four classes, in which the mouth (oral side) is oriented downward or forward and the anus (if it is developed as a separate opening) uppermost (or posteriorly) on the aboral side.

Asteroidea

The asteroids include all the starfish, of which there are over 1600 species. Each arm has an open groove, the ambulacrum, on its oral or downward side, extending from its tip to the centrally located mouth. There are two prominent rows of tube feet, one on each side of the ambulacral groove. The three orders into which the class is most commonly divided are the Phanerozonia, the Spinulosa, and the Forcipulata. The Phanerozonia and the Spinulosa include many common examples (e.g., *Astropecten, Henricia, Solaster,* and *Luidia*) that have more-or-less flattened bodies with two rows of marginal ossicles that give the animal a distinct and relatively stiff margin. In a few cases, the arms are very short, almost merging with the central disc so that the animal is almost pentagonal in outline. The Forcipulata are distinct from the others in that they have a relatively small central disc and long, nearly cylindrical arms. The marginal plates that outline and appear to stiffen the other two groups are less evident or missing entirely, giving the animals a more flexible appearance. Examples of forcipulates are *Asterias, Pisaster,* and *Pycnopodia.* This group is noted for numerous and prominent pedicellariae, which are either stalked or closely adhering, pincerlike elaborations that are distributed over the surface of the test.

Ophiuroidea

Ophiuroids include over 2000 species of brittle and basket stars. Owing to the snakelike flexibility and mobility of the arms of this class they are sometimes called "serpent stars." The digestive tract has a single opening on the downward side and does not extend into the arms. The arms are very slender over their entire length. The basket stars, e.g., *Gorgonocephalus*, are characterized by multiple branching of their arms at all levels from the base outward and in life often sit with their many arms extended outward and upward in the shape of a basket, and are thus able to entrap their prey. Typically, the long, slender arms of this group are very mobile, permitting relatively rapid movement and extensive coiling and entanglement. Basket stars are often quite large (up to 10 cm in disc diameter).

Echinoidea

The sea urchins, sand dollars, and heart urchins have no extensions comparable to the arms of asteroids or ophiuroids but are instead either slightly or markedly flattened ovoids. Their ossicles are closely interconnecting to form a rigid test. Echinoids are covered with spines ranging from 1 mm or less to several cm in length, depending on the species. Most urchins have a diverse array of highly developed and characteristic pedicellariae. The irregular forms have a definite anteroposterior axis superimposed on a pentaradial symmetry and normally move in the anterior direction with respect to this axis. Deep indentations or holes through the tests of some irregular sand dollars (cake urchins) are thought to facilitate their movements just beneath the surface of the sandy or muddy substrate. The oral side is downward and locomotion is achieved by means of wave-like movements of the spines and, particularly in the sea urchins, movements of the elongate tube feet. Examples of echinoid genera are *Echinus, Strongylocentrotus,* and *Dendraster.*

Holothuroidea

The sea cucumbers are generally either cylindrical shaped or flattened cylinders, although some bizarre forms, including one that resembles a free-swimming medusa (*Pelagothuria*) are known. The rows of tube feet (when present) and the ambulacral grooves are arranged along the length of the cyclinder, with three rows (the "trivium") grouped on the downward side and two (the "bivium") on the uppermost surface. A well-defined anteroposterior axis exists, with a tentacular ring around the mouth and an anus and internal respiratory tree at the posterior end. There are five orders, distinguished mainly by details of tube feet and tentacular morphology, and approximately 900 identified species. Virtually all are benthic dwellers, and most are found at depths less than 100 fathoms.

B. Evolutionary Relationships

Although echinoderms have a rich fossil record (600 million years), their relationship to invertebrates is not well understood. It has been commonly held that they derived from a dipleurula form, a crawling, bilateral animal with three pairs of coelomic sacs (Barnes, 1968). They are difficult to locate phylogenetically since the adults are not comparable to other animals. The free-swimming and bilateral larvae are taken to be indicative of ancestral states (Fell and Pawson, 1966). With the advent of sessile forms, bilateral symmetry was lost. The water-vascular system was derived from two of the sacs; the "pentactula" theory states that derivation was from a bilateral animal that had hollow food-catching tentacles around the mouth and that these tentacles became the radial canals of the water-vascular system.

III. SENSORY-MOTOR APPARATUS IN ECHINODERMS

A. Organization of the Nervous System

The nervous system of echinoderms shows no evidence of cephalization and very little evidence for aggregation of neurons into ganglia or centers that might be construed as executive or integrating headquarters (see Fig. 3). The overall nervous system has been described as a "republic of reflexes," implying that a considerable degree of peripheral autonomy exists. Evidence for this statement can be taken from the commonly made observation that mechanical disintegration of an individual animal into small fragments of arms, disc, test (echinoids), or body wall (holothuroids) produces fragments, each of which retains many, in fact, most, of the normal reflexes seen in that same structure in the whole animals. The integrity of these sensory-motor fragments extends down to the level of one or a very few tube feet, spines, or pedicellariae. While there is extensive and often dramatic evidence supporting considerable peripheral autonomy, it is also clear that a degree of integrative capacity exists, and in some cases, this capacity results in spatiotemporal coordination of surprisingly complex acts.[2]

The neuroanatomical basis for this extraordinary decentralization of sensory-motor control can be seen in the distribution and structure of sensory receptors and nerves in the epithelium and deeper body regions. Sources of sensory input are both discrete and diffuse. All tube feet have some mechanosensory or chemosensory capability. These capabilities are mediated by the nerve tracts, which encircle the extremity of each tube foot, pass along the length of the tube foot, and encircle its base. They are in turn, an integral part of the general basiepithelial nerve plexus, which lies within and just beneath the entire epithelium of the animal. While this plexus is made up of a diffuse nerve-net-like distribution of fine, very small unipolar and multipolar cells, there are also discrete tracts of nerve fibers passing more-or-less randomly over the surface. The density of these fibers is extremely high in places, and estimates are given as high as 70,000 per

[2]Moore (1910) objected to "reflexology" mechanistic interpretations of starfish behavior such as that described by Maier and Schnierla (1935): "They [Maier and Schnierla] assign to the activities of the tube feet the function of coordination and conclude that the central nervous system, consisting of the axial nerve tracts and the circumoral nerve ring, plays a minor role in coordinating the activities of the animal. They believe that coordination can be accounted for by the results of tension on the tube feet. In short, they believe that the starfish is carried about by its tube feet; the animal's central nervous system having little or nothing to do with it" (p. 314). Moore showed that severed nerve ring connections could affect coordination of movement.

mm² in the epithelium over the adambulacral region. In addition to serving the function of transmission of excitation from place to place over the body surface, it is clear that these fibers contain numerous primary sensory units or are, in general, directly excitable by chemicals, mechanical perturbations, or light.

Bullock (1965) has posed the question of whether the basiepithelial plexus qualifies as a true nerve net in the sense that excitation may be transmitted between distant points in the system around blocks or disruptions via alternative routes. Curiously, the two different responses studied (dermal papula retraction in asteroids and convergence of spines toward a

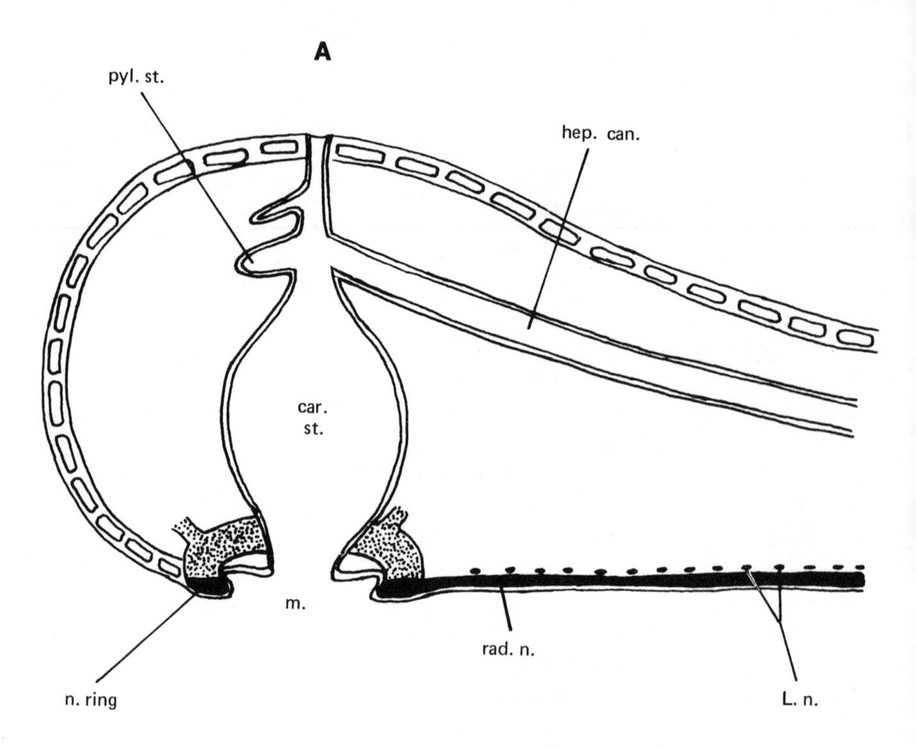

Fig. 3. Nervous system components in echinoderms. A, A medial-sagittal section through the body and one arm of a starfish. The nerve ring (*n. ring*) encircles the mouth area and sends branches (*rad. n.*, radial nerves) into the oral side of each arm. This component of the nervous system is considered to be an aggregation of basiepithelial plexus elements and to be primarily sensory in function. B, Cross section through a starfish arm. Lange's nerves (*L.n.*) occur as small aggregations of cell bodies distributed above each tube foot pair and are part of the hyponeural nervous system. They send processes into the walls of the bases of the tube feet and ampullae and into muscles that cross the ambulacral space, inserting on the most medial pair of ossicles; *ap.n.*, apical nerve; *car. stom.*, cardiac stomach; *hep. can.*, hepatic canal; *L.n.*,

local stimulus in echinoids) produced quite different results. Scratches made in the epithelium to cut through the nerve pathways completely blocked the convergence reflex, and no evidence was found that excitation could be passed around the scratch. This and other evidence suggested that conduction takes place via a system of straight lines between all points. Quite clearly, this suggests a form of nerve conduction not seen in other systems and quite unlike that expected of a nerve net in the usual sense of the term. On the other hand, reflex retraction of dermal papulae in asteroids showed evidence of passing around the ends of scratches in the basiepithelial plexus, implying a more usual form of nerve net capability.

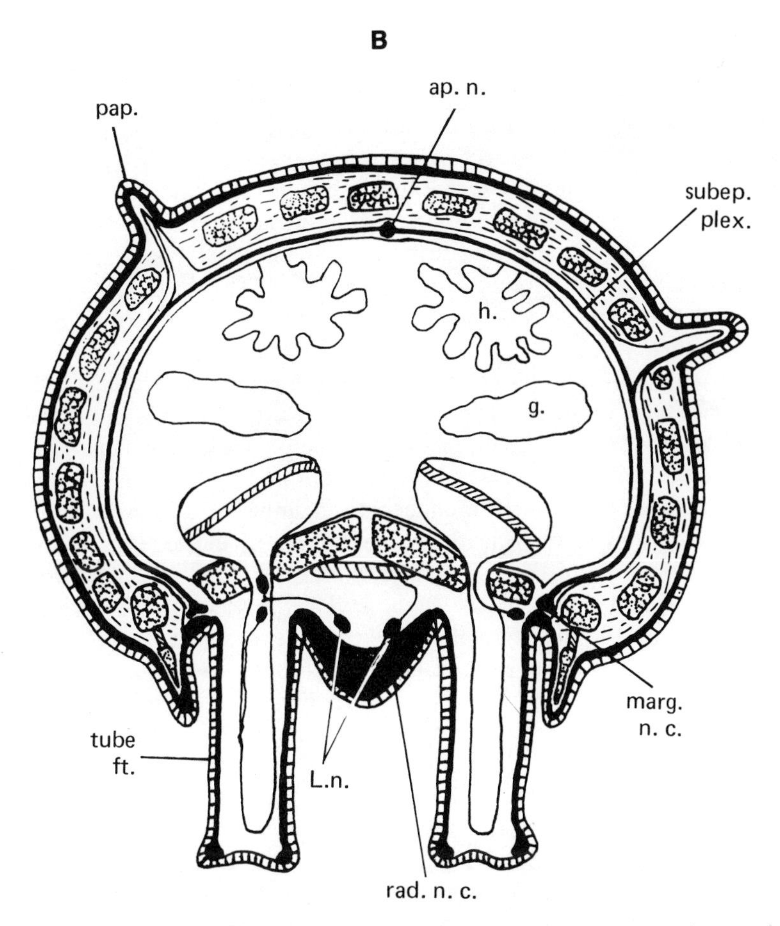

Lange's nerves: *marg. n.c.*, marginal nerve cord; *m.*, mouth; *n. ring*, circumoral nerve ring; *pap.*, papula; *pyl. stom.*, pyloric stomach; *rad. n.*, radial nerve; *rad. n.c.*, radial nerve cord in cross section; *subep. plex.*, subepithelial plexus.

In addition to the sensory capacities of the tube feet (perhaps used to palpate and "taste" the quality of food particles passing along the ambulacrum) and the entire epithelium, there are very few additional specialized sensory structures in echinoderms. Some tube feet, particularly the one at the end of each arm in asteroids and the several (usually an integral multiple of 5) arranged in a ring near the mouth in virtually all echinoderms, are known to have especially well-developed chemo- and tactile-sensory functions.

Four other prominent sensory structures occur in echinoderms. At the end of each arm in asteroids, there is one specialized tube foot (referred to above), which has a well-developed sensory innervation. At the base of each of these special tube feet there is a small, raised optic cushion, made of numerous pigmented ocelli, often with lenslike light-gathering accessories. These optic cushions have nerves originating in them that pass directly into and join the end of the radial nerve (see below). Both the anatomy and the physiology of these structures point to a primary light-sensing function. It should be pointed out, however, that these are not the only light-sensing structures in asteroids. In addition, similar structures do not occur at all in several other echinoderm classes. The basiepithelial plexus and perhaps also the radial nerve have direct light sensitivity (Takahashi, 1964), and diffuse light sensitivity probably occurs in all or most echinoderms with an exposed epithelium.

Prominent statocysts are not common in echinoderms. Generally, it is supposed that a sense of orientation is not well developed, and instead, animals tend to seek out situations in which contact with the substratum can be well established, regardless of the relationship to gravity. However, there are a number of relatively mobile sea cucumbers that have well-defined statocysts placed in or beneath the epithelium at the anterior end, where each radial nerve cord joins the circumoral nerve ring. Others are found along the radial cords of the bivium. The presence of these organs is consistent with the life style of these creatures, some of whom move relatively freely and even vigorously during escape reflexes. A sense of up–down may well be useful in the retaining or regaining of normal substrate contact under such circumstances and is quite obviously adaptive in burrowing species.

Holothuria show a broad-spectrum, nonspecific sensitivity to light and to mechanical and chemical stimuli applied to various parts of the body. It has been proposed that a "universal sense organ" (Olmsted, 1917) distributed over the glandulosensory warts and in pits on the tentacles may be responsible for this wide-ranging, relatively nonselective sensing ability. The ultrastructure of these organs includes evidence for the presence of ciliary and basal fiber structures that contact underlying nerve fibers

(Binyon, 1972). These sensory receptors have the elemental structures required for the range of sensitivity ascribed above.

The urchins also have numerous specially developed gravity receptors scattered over the surface of the test. These structures, usually in clusters or rows, are presumably modified spines and consist of a spherical globule atop a muscular and mobile column. The inside of the sphere contains a calcareous otolith, and the surrounding epithelium is ciliated. A nerve ring, again part of the basiepithelial plexus, surrounds the base of each sphaeridium.

B. The Ectoneural, Hyponeural, and Entoneural Systems

At the anatomical location where the basiepithelial plexus converges upon the oral opening, there is a confluence of the plexus with a well-defined nerve ring, which encirles the mouth. This nerve ring sends radial branches into each of the arms. The radial nerves pass superficially along the center of the ambulacral groove, between the two rows of tube feet. As well, there are lesser marginal nerve cords that pass laterally to the tube feet in each arm along their length. All along the length of the radial nerve, elements of the basiepithelial plexus from the walls of the tube feet and sides of the ambulacral groove send numerous fine processes into the radial and marginal nerve strands. This entire system (the oral ring and the radial extensions) is called the "ectoneural nervous system," although it is clear also that this system is validly thought of as an aggregation or concentrated transmission system for the basiepithelial plexus (Fig. 3). The ectoneural system is usually supposed to be primarily a sensory system, although there are no reasonable grounds to entirely preclude certain motor functions as well, particularly in regard to mediating aspects of control over tube foot movements and subepithelial muscles that bring about movements of arms, pedicellariae spines, or other body parts.

Two other nerve rings and concentrations of nerve bodies and processes occur in echinoderms, which interconnect and interact with the ectoneural nervous system. In asteroids, the two systems (hyponeural and entoneural) are virtually amalgamated and sometimes difficult to distinguish. They include nerve tracts that encircle the gut and rows of associated aggregations of nerve cells at the level of each pair of tube feet (called Lange's nerves). These fibers and nerves are supposed, on the basis of their anatomical connections, to be primarily motor in function. They (Lange's nerves) can be seen to send axons that innervate muscles connecting the two interambulacral plates (to draw the sides of the ambulacra together) and into the base of each tube foot. There they are presumed to act as motor centers, contributing via synapses to motor control of the tube feet both by fibers that pass up and innervate the ampullae, which control the water pressure of the tube foot (and hence extension or retraction), and other fibers, whose processes pass down into the walls of the tube feet, where they are supposed to directly innervate the tube foot musculature and thus control the direction and extent of bending. These nervous components are not so well defined nor continuous as is the ectoneural system described above. Each group of Lange's nerves at the bases of the tube feet has some of the features of an autonomous ganglion containing approximately 20 neurons, although the level of organization hardly warrants even the term "ganglion."

In crinoids, the situation is somewhat different in that the oral side of the animal is upward. Corresponding to this reversal of orientation, the relative prominence of ectoneural, hyponeural, and entoneural systems is altered significantly. The entoneural (primarily motor) system is more extensively developed, while the ectoneural sensory system is much less well defined. The entoneural tract encircles the chambered organ of the basal calyx and sends a branch into each arm.

In each case in the echinoderms, it is the neural ring and branches on the downward side (whether oral or aboral) of the creature that predominate, both in extent of organization and in significance to the animal in terms of sensory-motor control. This observation may reflect the obvious fact that, in every case, it is the side of the animal in contact with the substratum that first encounters stimuli relevant to locomotor or feeding behavior. Clearly, the irregular urchins and the holothurians, which have an anteroposterior axis as well, must be considered special cases. Again, however, the most highly developed portions of the nervous system in these forms are the parts that first encounter stimuli, namely, the most anterior regions.

C. General Behavior

Time Dimensions

Normally we are accustomed to animals that are doing several things, some simultaneously, but certainly displaying different behaviors within seconds or minutes. Such variability keeps the experimenter interested during long hours of data collection. However, among the echinoderms it is frequently impossible to detect life—some holothurians have been observed to remain immobile for two years and the sand dollar can manage a top speed of 14 mm per min; these animals would not make particularly good candidates for T-maze learning. Some sample rates of movement among

Table I. Reaction Times of Echinoderms to Various Stimuli (Warden *et al.*, 1940)

Species	Investigator	Stimulus	Part stimulated	Time in sec	Range
Holothuria surinamensis	Crozier (1915)	Light contact	Tentacles	0.3	
"	"	" "	Anterior surface	0.4	
"	"	" "	Cloaca	0.6	
"	"	Sudden shading	Tentacles	1.2	0.4–2.8
"	"	0.5 cc KCl (M/10)	Tentacles	0.48	0.2–0.9
"	"	0.5 cc KCl (M/10)	Mid-body	3.2	2.0–5.0
Asterias	Moore (1921)	Contact	Side of ray	2.8	
"	"	White light 26,000 CMS	Tube feet	1.5	
"	"	White light weak	Tube feet	3.0	
Centrostephanus longispinus	v. Uexküll (1897)	Shaking	Entire body	0.2	0.11–0.28
"	"	Sudden shading	Entire body	0.65	0.54–0.85
"	"	Partial shading		0.5	0.47–0.52

various species are the following (Warden *et al.*, 1940):

Asterias	25.0 mm/min
Echinarachnius parma	13.7 mm/min
Holothuria surinamensis	2.5 mm/min
Stichopus panimensis	67 mm/min
Comatella	85 mm/min

There are a few exceptions: *Pycnopodia helianthoides*, for instance, has been clocked at 1500 mm per min (Mauzy *et al.*, 1968). Reaction times, although showing considerable variability, are not quite as impossible (Warden *et al.*, 1940) (see Table I).

Orientation and Associative Behavior

Although many echinoderms lack statocysts, some still display a negative "geotropism" and can usually be found climbing up vertical walls or objects in laboratory aquaria. It has been assumed that the weight pull on the attached tube feet acts as the cue for those lacking statocysts (Fraenkel and Gunn, 1961). The burrowing behavior that is common among the sand dollars can be used to differentiate some species (Reese, 1966). The tendency of many species to cover themselves with objects could be due to a requirement for added weight in order to better resist turbulence, to camouflage themselves, or to avoid light (Reese, 1966). The echinoids and ophiuroids in particular show negative responses to light, reactions that could be more extensively explored in behavioral modification studies. In the asteroid *Asterias forbesi* there is a daily rhythm of positive phototaxis (Rockstein and Spritzer, 1960) with the maximal sensitivity to light in this animal at 450 nm.

There is some specificity in predation, another characteristic that might be used in explorations of plasticity—the asteroid *Pisaster brevispinus* causes burrowing in *Dendraster,* while *Astropecten armatus* does not (Reese, 1966). Aggregation among members can also be observed, but this is probably a substitute for something in the environment; Allee (1958) noted that brittle stars are seldom seen in aggregation in natural situations, yet when they were placed in a pail of just seawater they clumped together. In clean aquaria, they also aggregated, but if eel grass was added, no aggregations were observed. To further test this, Allee put twisted glass rods in a clean aquarium and found that the brittle stars remained isolated; removal of the materials produced aggregation.

Although the phenomenon of symbiotic activities in echinoderms is not particularly pertinent to issues of their plasticity, these relationships provide some interesting possibilities for observation of behavioral preferences and changes in *other* phyla. Davenport (1966) has presented an interesting treatment of symbiosis involving echinoderms. Echinoderms offer some particular advantages in associations: they live in detritus-free zones where oxygenation and plankton are adequate, their ciliated surfaces prevent stagnation, the various surface structures provide good protection, and they have few natural predators. Davenport provides some excellent examples of such relationships. The fish, *Aeoliscus strigatus,* for example, orients so that its stripe is parallel to spines of its urchin host. A parasitic fish is known to enter the cloaca of the sea cucumber and eat the gonads and branchial tree.

The commensals have apparently evolved complex signal detector mechanisms for orienting toward their echinoderm hosts. Davenport calls for the application of ethological principles in the analysis of the sign stimuli and designed an "olfactometer" that tests the specificity of reaction (see Fig. 4). Worms, for example, can choose either of two compartments, and the apparatus is designed so that water can flow from either compartment. The various connections permit randomization of substances given to the worm. Either whole animals or specific extracts of hosts can be placed in the compartments.

Davenport's studies showed that 90% of the commensals entered the arm containing host factors. The host-attracting factors were not stable and lost their potency after 24 hr. Damage to the host frequently inhibited an approach. Some commensals were not attracted by factors released into the medium and required actual contact. In reviewing the question of whether echinoderms actively seek such associations, Davenport concluded that it is rare except in Ophiuroidea.

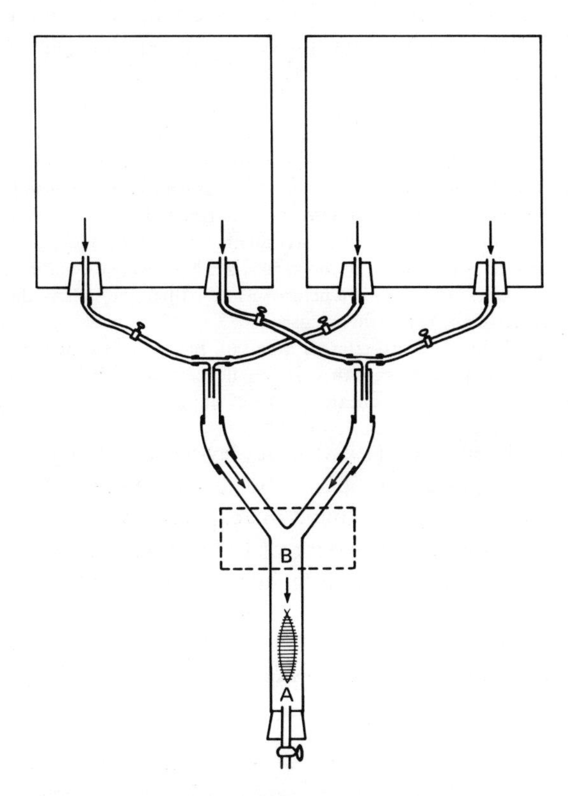

Fig. 4. Plan of Y-tube experimental apparatus (olfactometer). The cross connections permit the randomization of effluents reaching the test animal (*B*). Flow is initiated by the opening of the stopcock at *A*. (Davenport, 1966; reproduced by permission.)

Pedicellariae Characteristics

It might be useful and more productive if investigations focused upon the behavior of the pedicellariae of echinoderms. Jennings (1907) provided a somewhat provocative discussion of some of their functions and wrote of these organs as if they were separate little animals, discussing several properties that should be of interest to the behavioral biologist.

The major actions of the pedicellariae involve attachment (they rise to the top of a spine and seize an object) and withdrawal (closure). With repeated stimulation of the spines by means of a bristle, Jennings found that he could get the pedicellariae to rise (facilitation), whereas slow pressure on the spines did not elicit a response. Repeated stimulation of the gills near pedicellariae resulted in response decrements for a few strokes followed by their opening. The responses of the pedicellariae can be used to assess input integration capacities that involve more than mere nerve net conduction. The following passage provides a good example of the possibilities:

> What distant rosettes (clumps of pedicellariae) shall rise when a given point on the body is stimulated depends more on the recent history of the different parts than on the permanent anatomical connections. . . . In a quiet starfish a certain region (say, the tip of the arm a) is stimulated, till the rosettes of that region rise to the attack. These are allowed to subside, then a few moments later another part of the body, at a distance (say, the tip of the arm d) is stimulated in the same way. Now its pedicellariae rise, and *also the pedicellariae of the part stimulated* before (near the tip of the arm a), though the pedicellariae of the intervening region do not rise.
>
> Thus the reaction of the rosettes in a certain region leaves them for a time in a changed physiological state, so that they are readier to react to impulses coming from a distance than are even the rosettes nearer the point stimulated (p. 81).

Campbell and Laverack (1968) have devised more objective methods of observing pedicellariae responses by using photoelectric recording methods. The many suggestions found in Jennings's work will hopefully be replicated and extended; the behavior of the pedicellariae may be a more sensitive key in questions of echinoderm integration capacities and plasticity.

IV. LEARNING DEMONSTRATIONS

Generally, learning experiments have dwelled upon improvement in the speed and nature of the righting reflex and escape learning, with only three attempts at associative conditioning. The existing literature is characterized by its sparseness and by the lack of any systematic, well-controlled demonstration.

A. Righting Reflex and Persistence of Movement

Jennings noted in 1907 that starfish may show a "persistence" of movement. For example when an animal was moving away from the sunlight and then turned around so that it was headed toward the sun, it continued in that direction for 5–7 sec. Also, if an animal was given a solid object to touch and the object was then removed, movements continued as if the object were there. Such anecdotal observations are interesting with respect to neural properties that might subserve the persisting behavioral activities. Echinoderms are not noted for rapid and systematically changing behavioral repertoires, and general behavioral modes that persist over time may be adequate in their normal environments:

> . . . time is a relatively unimportant factor in the life of these animals and . . . therefore the biologically significant fact is the successful completion of a given act no matter how varied or time consuming the movements are which lead to its accomplishment (Warden *et al.*, 1940, p. 438).

For the most part, interest in response persistence and change has centered on the righting reflex of the starfish. As Jennings stressed, this reaction is important since activities such as feeding occur with the central side down. When overturned, the animals cannot move and there is the possibility of gills' being damaged on the substrate. Jennings concluded that there is some persistence in the major mode of sighting. If a starfish used a particular pair of rays to right itself in one condition and was then transferred to different circumstances, the same pair of rays was still used. Jennings believed that the characteristics of previous reactions were generally repeated and presents some data that tend to substantiate this in some respects. For example, in two animals the following data were obtained on persistence of ray usage in righting:

Ray	a	b	c	d	e
Subject A:					
Frequency of initial test	15	1	4	3	17
3 days later	3	1	1	8	8
Subject B:					
Frequency of initial test	15	1	0	5	14
3 days later	7	3	3	2	4

Such data leave the modern investigator with an uneasy feeling but they certainly are suggestive of experiments that would assess the persistence and hierarchical organization of particular reflexes over extended periods of time.

The same uneasiness results from Jennings' description of his attempts to modify the righting behavior of the starfish. Extensive training was actually conducted in only two animals. The reasons given are not unlike those occasionally used by present-day investigators:

> The small number of individuals employed in extended and thorough work was due to the very great amount of time required in the experiments, and to the fact that the successful method of work was not hit upon till near the end of the season. A given starfish may require two to seven minutes for its reaction, and the careful description to be written requires still more time. Then there is the period of rest between successive lessons. Altogether, by the time five specimens have been given ten lessons, at intervals, the part of the day left for other work is small. Moreover, the work is extremely tedious (Jennings, 1907, p. 156).

Today's graduate students' complaints are not dissimilar.

Jennings's basic strategy was first to determine the natural ray preference in righting. In some cases one ray clearly predominated, while in others pairs of rays were used. He attempted modification of ray preference by placing the animal on its back and, manually or with forceps, preventing it from using any rays to right itself save the least preferred. In one subject ("Starfish A"), 30 tests were run, and it was determined that the strongest predilection was to use ray *a* 25 times and $a + e$ 15 times. Rays *b* (twice) and *c* (5 times) were used less and never in combination. At first, ray *b* was never used and the starfish persisted in using other rays. An accidental contact with the substrate on the fifth trial caused *b* to be used, and Jennings prevented all other rays from assisting. However, if *b* had made contact and the other rays were released, the dominant rays ($a + b$) took over and executed the turning. What was accomplished during the first 10 trials was the establishment of ray *b*'s ability to right the animal *as long as other rays were prevented from involvement.* On the second day of training (8 additional trials), attempts were made to induce cooperation between *b* and *c,* but adaptive coordination between these two rays was impossible at this point. It was observed that rays $b + a$ worked well together. After 18 trials, *b* was actively involved in the righting reflex, and during 5 test trials in which there was no experimental manipulation, *b* was one of the first rays to take hold and lead the righting reflex. Attempts to induce coordinated activity between *b* and *c* were never fully successful, although some modification was observed. After 180 trials, and a two-day interval, test trials showed the following frequencies: $b + c$, 4; $c + d$, 3; $a + e$, 2; $d + c$, 1. A further five-day rest resulted in the almost complete disappearance of the modified preference for ray *b* usage. These observations are interesting and strongly hint at transient reflex modification and moderate persistence of the new response.

Ohshima (1940) noted a reduction in righting time with practice in the sea star *Oreaster nodosus.* The range of righting time was quite wide, varying from 2 min 20 sec to 1 hr and 30 min for the "folding" response

Table II. Righting Latencies in *Oreaster nodosus* (Ohshima, 1940)

Number of individual	Trial					
	1st	2nd	3rd	4th	5th	6th
13	2^m 50^s	2^m 20^s	2^m 35^s	2^m 40^s	3^m 30^s[a]	7^m 45^s[a]
5	15 30	12 00	13 45	14 10[a]	1^h [a]	
14	7 15	5 40	6 15	6 30	10^m 45^s[a]	
9	7 00	6 00	5 00	9 30	2^h 30^m[a]	
1	6 00	5 30	5 30	4 30	[b]	
16	10 00	5 00	3 50	5 45		
2	10 00	7 00	9 10			
7	12 15	7 00	10 30			
8	6 00	4 30	8 00			
15	9 30	6 15	6 45			
12	1^h 27^m	7 30	7 00[a]			

[a] The nerve or nerves severed.
[b] Died at last, being unable to right itself.

and 3 min 20 sec to 2 hr for a "somersault." Ohshima's data are summarized in Table II and demonstrate a drop in righting time during the first three trials, followed by an increase that, according to Ohshima, is due to fatigue. Severing the dorsal or ventral radial nerves had only a slight effect on the ability and speed of the reflex. Ohshima concluded that considering the sea star's habitat and behavior on the sea floor, it is unlikely that righting movements could be used much.

Moore, in 1910, took issue with Jennings's conclusions concerning the modifiability of the righting reflex. According to Moore, it is not physiologically damaging for the animal to be upside down; for example, animals frequently extend their arms out on the underside of the surface and they will remain half upside down in this position. This behavior suggests that there is no damaging displacement of internal organs in this position. With respect to the righting reflex modification, Moore suggested that the arms that initially contact the substrate lead the righting reflex and that inhibiting impulses are sent to the opposite arms. Active arms were treated with acid or stimulated with a glass arm, and Moore found that previously inactive arms were then used. He attributed the findings of Jennings to inhibitors resulting from injury to the tube feet of active arms. Kjerschow-Agersborg (1918) found no improvement in righting reaction with practice in *Pycnopodia helianthoides;* negative results were also obtained by Glaser (1907) in the brittle star *Ophiura brevispina.*

Persistence of movement direction has been reported in brittle stars (Cowles, 1910). One arm was permitted to contact a surface and then the animal was pulled away (but left in the same orientation). The brittle star returned to contact the surface and was not deterred by a strong light. Position memory was also suggested in experiments in which animals were removed from one place and put in an inverted orientation at an opposite corner; they returned to the original position. It has been suggested by others that these results are due to the stimulating effects of breaking contact with a surface, which merely trigger off movement in particular directions (Warden *et al.*, 1940).

B. Escape Behavior

The work of Preyer (1887) on ophiuroids in the last century was instrumental in producing much investigation and controversy over the possibility of escape learning. Preyer put a rubber tube over one arm of a brittle star and noted the latency and characteristics of attempts to remove the tube. His observations led him to conclude that there was both a decrease in escape time and reduction in ineffectual movements. Glaser (1907) took issue with Preyer's conclusion that "intelligence" had been demonstrated. Carefully documenting the major types of movement in *O. brevispina*, he found that there was considerable variability in character over time and that there were seasonal differences. When a tube was placed over an arm, the animal rarely moved in that direction. At first, there was a latent period during which the animal was motionless, followed by crawling activity in which the arm was dragged, and finally a period of writhing and arm waving. Some reflexes were more effective than others, but Glaser did not observe a decrease in escape time, nor was there an increase in the use of the most effective reactions. The latencies of five subjects are included in Table III. Glaser noted that ineffective movements are as likely to be repeated as effective ones and discusses the possibility that since ophiuroids have such a wide variety of responses and since both effectual and ineffectual responses can occur simultaneously, there is the possibility of conflicting responses and, thus, little improvement over time.

Ven's (1921) analysis of the escape behavior of starfish remains one of the more interesting and convincing earlier research papers. Evidence was obtained that suggested a more "goal-directed" type of learning as well as improvement of escape responses with practice. The apparatus used throughout the experiment is portrayed in Fig. 5. The animal was placed on

Table III. Escape Latencies in *Ophiura brevispina* (Glaser, 1907)

	Trials[a]						
Individual	1	2	3	4	5	6	7
A	3′45″	5′00″	2′00″	3′00″	6′00″		
B	0′30″	4′30″	1′00″	1′1″	4′00″		
C	0′45″	1′45″	0′1″	5′00″	1′30″		
D	2′00″	3′00″	7′30″	3′00″	3′00″	2′15″	3′30″
E	1′30″	0′45″	1′30″	1′40″	5′15″	3′30″	

[a] Measurements include the latent periods.

the bed of clay and pegs were inserted between the arms; a restraining collar or bar was placed over the two most active arms (usually the ones adjoining the madreporite). Both the duration of the escape trial and the characteristics of the escape attempts were noted. To begin a trial, the plate was lowered into the bottom of a tub, and the position of the plate was rotated from trial to trial to prevent the animal from forming associations and responses with extraneous stimuli. At first, the results were negative—three subjects actually showed an increase in escape time in successive trials—and Ven offered an interesting interpretation of this result: after a trial the animal was removed from the plate and returned to the aquarium, a procedure that could be considered a form of punishment since damage to the tube feet was likely. Additionally, the environment of the tub was new and presumably added to the general excitatory state of the animal. These considerations are indicative of a certain degree of sensitivity in Ven to some experimental factors that others have recently uncovered in invertebrates. For example, similar conclusions were reached in maze learning in the planarian and in pond snails. To avoid this negative reinforcement possibility, Ven ran all subsequent experiments with the plate lowered in the animal's aquarium, and escape from the plate achieved what was probably a more positively reinforcing situation. The general training procedure was also modified somewhat so that it consisted of two stages: in stage I, the pegs were omitted and only the collars were used to prevent movement in the direction of the two most active arms; in stage II both the pegs and the collars were used. With these changes, decreases in escape time and increases in effective movements were observed. For example, in one subject, out of 120 trials, 68 eventually led to escape and this was accompanied by a decrease in escape time. The persistence of ineffective movements was observed to drop—a representative curve is shown in Fig.

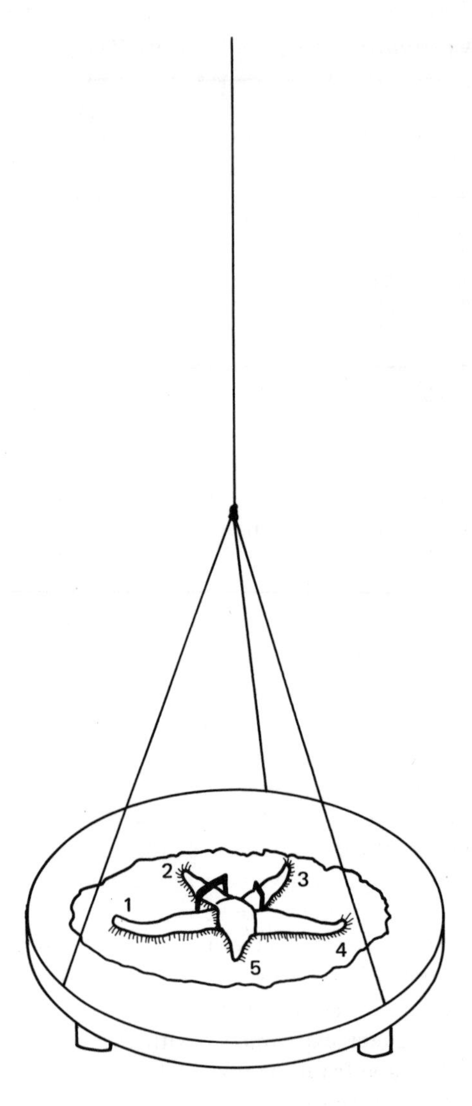

Fig. 5. The arrangement devised by Ven (1921) to study the ability of the starfish to escape. The plate was lowered into the tank so that successful escape movements returned the animal to a familiar environment.

6. It was possible for one of the better subjects to show a drop of 30 to 10.5 min in escape time after only 10 trials. Interestingly, with continued training, Ven was able to present the subjects with more trials per day. At first, it was necessary to interpose several days between trials, but later, up to 6 trials per day with a 30-min intertrial interval could be given. A day off (Sunday, of course) was always followed by a transient decrement in performance on Monday (see Fig. 6). A negative effect was produced when

too many trials were given in one day. Ven noted that he frequently had to poke animals when they were inactive, and because of this, latency measures were not the most accurate indication of the rate of learning. Poking was necessary in the first few trials and led to a decreased escape duration—if the animals had been left alone, an even steeper curve would have been obtained.

At the completion of training, Ven ran what were called "control trials." In these trials, the collars were removed but the pegs were left. *Subjects continued to escape in the previous direction even though the more active arms were now free.* In one subject, the dominant arms were used only once in 21 control trials to effect escape. According to Ven, this alteration in escape direction was not due to a suppression of activity in the active arms; *it appeared that these arms were actually as active but were used in an integrated fashion with the remaining arms to achieve a new direction of movement.*

Recent attempts at demonstrating escape learning in *A. forbesi* have been conducted by Corson (1974). Because the earlier work of Jennings

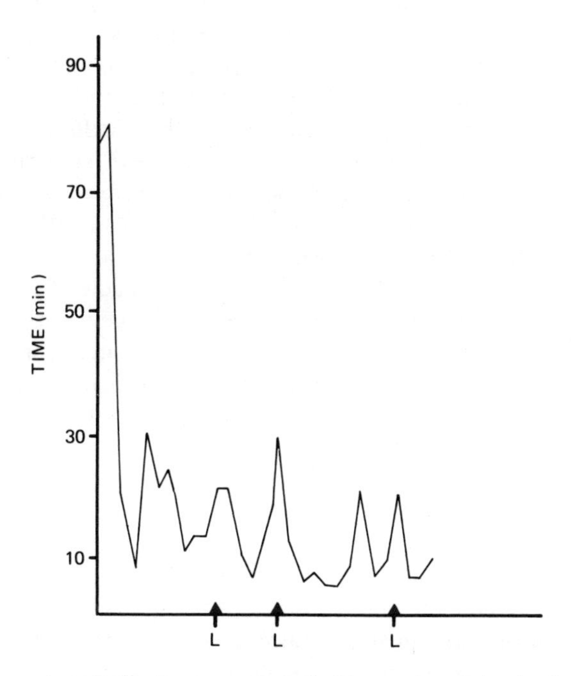

Fig. 6. Time spent on ineffective movements during escape training in the starfish (Ven, 1921). *L* indicates a Monday on which training was resumed after one day of rest. (Ven did not label this abscissa any more than indicated.)

(1907) had been criticized on the possibility of damage or sensitization of the rays, the following procedures were adopted: The arms were numbered clockwise from the madreporite. Subjects were divided into five groups, each of which had a particular arm free on the first trial. These groups were subdivided into groups that had their subsequent trials on the same arm; on arms in clockwise order from the starting arm; on arms counter-clockwise from the starting arm; and a mixed group that had an arm free in random order on subsequent trials. The results of this study are not easily interpreted. Over 80% of the animals showed some improvement over the first five trials (with a 2-min intertrial interval), but when continued trials were presented, the animals slowed down. With the exception of a group with a 10-min intertrial interval, lengthening the interval did not seem to alter this pattern. Rotating the arm and varying the trial interval were also without consistent effect.

C. Associative Learning

One of the more extensive attempts at establishing associative conditioning was that of Emil Diebschlag in 1938. Working with *Astropecten, Ophiothrix,* and *Psammechinus,* Diebschlag examined the capacity of animals to form associations between surface textures, light and dark, and combinations of these stimuli. The experimental chamber consisted of a Petri dish 15 cm in diameter. A 6-cm strip in the middle of this dish provided a differential stimulus substrate. In the rough/smooth discriminations, this central strip was smooth and the sides were roughened by being coated with emery. The animal was positioned on the smooth portion of the Petri dish to initiate a trial, and when one-third of an arm had extended into this area, punishment (touch) was given in the middle of a group of pedicellariae. Diebschlag noted that it was necessary to deliver the punishment later than the conditioned stimulus or CS (rough contact) in order to establish an effective association. Additionally, the application of a too-intense unconditioned stimulus, notably in *Ophiotrix fragilis,* resulted in such hyperexcitement that purposeful behavior was impossible. At proper intensity levels, the animals moved from the rough to the smooth area quite easily, a reaction that frequently carried them to the rough surface on the other side of the strip, where they were again punished. The results of two sample subjects are summarized in Table IVA. The training of Echinoida in this paradigm proved to be more difficult, as the spines interfered with exact stimulation and the experimental chamber was not ideal for spherical animals. Reversing the situation and placing a rough strip in the middle of the Petri dish as a safe area and punishing for lateral movements onto a smooth surface did not yield much evidence of learning.

In experiments in which a wavy/smooth surface discrimination was tried, a different experimental chamber was used. A glass rectangular aquarium was constructed with one half made of smooth and the other half of wavy glass. The animal was shocked when extension was made into the side with the smooth surfaces. Results were poor for *Astropecten* and moderate for *Ophiotrix*. Examples of subjects' responses are presented in Table IVB. Reversing the contingencies and punishing extensions onto the wavy side yielded good avoidance in both *Astropecten* and *Ophiotrix* (see Table IVC). Discriminations between two different textures in *Astropecten* (wavy versus rough) also demonstrated success; the example provided by Diebschlag shows avoidance after three punishments but reversal of the discrimination was not successful.

Experiments with light/dark discriminations in *Astropecten* were also mainly successful. Light was directed at the animals from beneath, and a dark area was produced by the positioning of a strip of opaque paper on the bottom. As in the rough/smooth discriminations, a round Petri dish was used with the central strip constituting the "safe" area and punishment delivered in the lateral areas. Light avoidance was achieved and the last 36 contact choices are presented in Table IVD for one subject. Diebschlag commented on the objection that punishment merely induced a state of photonegativity by pointing out that under his laboratory conditions, the natural trend of his subjects was to become more photopositive. This is, of course, not a particularly strong point since Diebschlag never ran what would be termed "sensitization" controls, i.e., the random presentation of punishment and light. However, what does lend some support to Diebschlag's conclusion that this type of associative conditioning is possible were his findings with respect to dark avoidance conditioning. In these experiments, the dish was dark except for a 6-cm-wide strip down the middle. As in previous conditions, the animal was placed on the central strip and the contacts with the lateral portions were recorded. Only animals that were neutral toward light could be used in these experiments, since the photopositivity of animals that were kept in the laboratory could account for any preference and also because animals that were photonegative initially could not be changed. An example of a subject given this type of training is given in Table IVE. The use of various combinations of stimuli (dark/rough versus smooth/light) also yielded positive results, but what is interesting about these attempts is the finding that compounding stimuli did not improve performance over that of simple discriminations.

Evidence of tactile conditioning in *Asterias rubens* has been obtained by Sokolov (1961). Previous research had demonstrated the possibility of light conditioning in the starfish by the use of food reinforcement. These analyses were extended by examination of whether the tactile information

Table IV. Data Examples (Diebschlag, 1938)[a]

Contact with border	*Astropecten*	*Ophiothrix*	*Psammechinus*	Contact with border	*Astropecten*	*Ophiothrix*	*Psammechinus*
A. Examples of training results using rough/smooth discriminations (rough surface contact punished)							
1	P(1,5)	P-(2)	P	21	P-(4)	P-(3)	P
2	P(2,3)	P-(4)	P	22	A-(1)	A-(2,5)	P
3	P(1,5)	P-(1,5)	P	23	A-(1)	A-(2)	P
4	P(3)	A-(3)	P	24	A-(1)	A-(2)	P
5	P(3)	A-(3)	P	25	A-(1)	A-(5)	P
6	A-(1)	P-(3)	P	26	A-(1)	A-(5)	P
7	P-(1)	P-(3)	P	27	A-(1)	A-(5)	P
8	P-(3)	P-(3)	P	28	A-(1)	P-(1)	P
9	P-(1)	P-(5)	P	29	A-(2)	A-(1)	P
10	P-(4)	A	P	30	A-(2)	A-(1)	P
11	P-(1,2)	P-(2)	A	31	A-(2)	A-(1)	A
12	P-(2,3)	P-(5)	P	32	P-(2,3)	A-(1)	P
13	P-(2,3)	A(5)	A	33	A-(2,3)	P-(5)	A
14	P-(2,3)	A(5)	P	34	P-(2,3)	P-(5)	P
15	P-(5)	A(5)	A	35	P-(5)	P-(1,5)	A
16	A-(2,3)	P-(5)	P	36	P-(5)	P-(1,5)	P
17	A-(2,3)	P-(4,5)	P	37	P-(4)	P-(2)	P
18	A-(2,3)	P-(3)		38	P-(1)	A-(3)	
19	P-(4)	P-(5)		39	P-(3,4)	A-(2)	
20	P-(2)	P-(3,4)		40	A-(1,2)	A-(3)	

Contact Number	*Astropecten*	*Ophiothrix*	Contact Number	*Astropecten*	*Ophiothrix*
B. Examples of avoidance of a smooth surface in two subjects					
1	P-(1,5)	P-(5)	17		A-(3)
2	P-(1)	P-(5)	18		P-(4)
3	P-(1,2)	P-(5)	19		A-(2)
4	P-(1,5)	P-(1)	20		P-(1,5)
5	P-(3)	P-(3)	21		P-(4,5)
6	P-(2)	P-(4)	22		A-(3)
7	P-(2)	A-(4)	23		P-(1,3,5)
8	P-(4)	A-(5)	24		P-(4)
9	P-(5,1)	A-(5)	25		P-(4,5)
10	P-(1)	A-(5)	26		P-(5)
11	P-(2)	A-(5)	27		P-(3)
12	P-(4)	P-(2)	28		A-(1)
13	P-(2)	A-(4)	29		P-(3,5)
14	P-(5)	A-(5)	30		P-(5)
15		P-(4,5)	31		P-(5)
16		P-(3)	32		A-(5)

Table IV. (Continued)

Contact Number	*Astropecten*	*Ophiothrix*	Contact Number	*Astropecten*	*Ophiothrix*
C. Examples of avoidance of wavy surface in *Astropecten* and *Ophiothrix*					
1	P-(1,5)	P-(5)	19	P-(3,1)	P-(1,2)
2	P-(3,4)	P-(5)	20		P-(3)
3	P-(4,5)	P-(1)	21		P-(3)
4	P-(1,2)	P-(5)	22		P-(1,5)
5	P-(5,1)	P-(4)	23		P-(5)
6	A-(5)	A-(4)	24		A-(1)
7	A-(5)	P-(5)	25		A-(1)
8	P-(3,4)	P-(3,4)	26		P-(1,5)
9	A-(4)	P-(2,3)	27		P-(2,3)
10	A-(4)	P-(2)	28		A-(2)
11	A-(4)	P-(2,3)	29		A-(2)
12	P-(4)	P-(2)	30		P-(3)
13	A-(5)	P-(1,2)	31		A-(4)
14	A-(4)	A-(1)	32		P-(4)
15	A-(4)	P-(2)	33		A-(2)
16	P-(3,4)	P-(1,5)	34		A-(2)
17	A-(1)	P-(2,3)	35		P-(3)
18	A-(1)	P-(4)	36		P-(2,3)

Contact Number	*Astropecten*	Contact Number	*Astropecten*
D. Examples of light avoidance in *Astropecten*			
20[b]	P-(5)	38	P-(5)
21	P-(3)	39	P-(2,3)
22	P-(1)	40	P-(5)
23	P-(1)	41	A-(4)
24	P-(3)	42	A-(1)
25	P-(1)	43	A-(5)
26	P-(4)	44	P-(3)
27	P-(3)	45	A-(4)
28	P-(4)	46	A-(4)
29	A-(4)	47	A-(5)
30	A-(3)	48	A-(2)
31	P-(3,4)	49	P-(1)
32	P-(3)	50	A-(1)
33	P-(2)	51	A-(5)
34	P-(2)	52	A-(2)
35	P-(5)	53	P-(4)
36	P-(4)	54	A-(2)
37	P-(2)	55	A-(4)

Table IV. Data Examples (Continued)

Contact Number	*Astropecten*	Contact Number	*Astropecten*
	E. Examples of dark avoidance learning in *Astropecten*		
1	P-(1)	19	P-(2,3)
2	P-(1)	20	A-(5)
3	P-(3)	21	A-(5)
4	P-(1)	22	A-(1)
5	A-(4,5)	23	A-(5)
6	A-(4)	24	A-(4)
7	A-(4)	25	A-(3)
8	P-(4,5)	26	A-(3)
9	P-(1)	27	A-(4)
10	P-(3)	28	A-(2,3)
11	P-(1)	29	A-(2)
12	P-(5)	30	A-(3,4)
13	A-(4)	31	P-(4)
14	A-(4)	32	A-(4)
15	P-(3)	33	A-(3)
16	P-(2)	34	A-(2)
17	A-(1,2)	35	A-(2)
18	A-(1)	36	A-(4)

[a] Numbers in parentheses refer to the arms that initiated movement.
[b] The first 19 contacts were all punished.

obtained from the substrate could be used as a signal to determine the direction of movement.

All animals were placed in the aquarium depicted in Fig. 7. A plate, containing numerous wooden pegs, was placed on the floor of one of two compartments in the aquarium. This pegged substrate served as the conditioned stimulus and was randomly alternated between the compartments. The unconditioned stimulus consisted of half a mussel. At the beginning of a trial, the starfish was placed in the front, noncompartmentalized portion of the aquarium so that the center of the body coincided with the border of the smooth and rough surfaces; the arms were also laid over both surfaces. The food reinforcement was given for 10 min in most cases at 5–15 cm from the entry point of a compartment.

Prior to training, ten control trials were given to each starfish. On each trial, the starfish was placed in the starting area and allowed to move freely. The preferences are summarized in Table V and demonstrate that four subjects preferred the rough, three the smooth, and two showed no

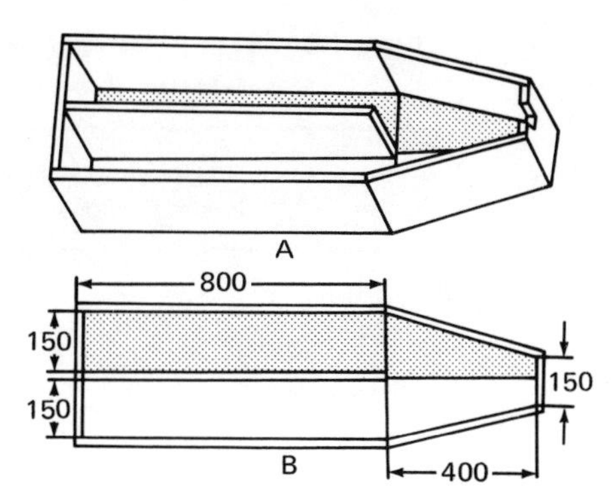

Fig. 7. The two-choice apparatus of Sokolov (1961). The rough surface is indicated by dots.

preference. After completion of the control experiments, training was initiated. The food reinforcement was given in association with the *least preferred compartment.* Thus, subject numbers 15, 16, 17, 37, and 48 were reinforced with food upon entering the compartment with the smooth floor; subjects 26, 46, 47 and 49 were reinforced when selecting the rough floor.

When compared to the control data, the surface preferences were changed in all subjects. The data summarized in Table VI show that the first correct choice of the surface occurred after 4–9 trials and that the response "stabilized" at 12–13 trials. Subjects 25, 16, 17, 26, and 37 were

Table V. Preference of Starfish for Smooth and Rough Surfaces in Control Experiments (Sokolov, 1961)

Subject number	Percent time in	
	Smooth	Rough
15	20	70
16	10	90
17	20	80
26	60	40
37	30	60
46	50	50
47	60	40
48	50	50
49	90	10

Table VI. Conditioned-Reflex Development in *Asterias rubens* (After Sokolov, 1961)

Subject #	Trial at which first CR appeared	Trial at which stabilization occurred
15	7	13
16	6	12
17	7	17
26	9	23
37	4	23
46	6	15
47	6	18
48	6	20
49	6	19

inadvertently transferred to aquarium conditions in which the temperature was much higher. This factor appeared to induce a greater variability in the time of formation and in the stability of the conditioning.

Sokolov included some suggestive data concerning long-term retention of the choice. The training of subject 15 was interrupted for three months after it had displayed eight out of ten correct responses to the smooth side. When training was renewed, only one trial was needed to reestablish the conditioning. Although the overall preference for the smooth surface dropped to 40%, the subject never entered the rough compartment. Similar retention is also reported for another subject.

Landenberger (1966) has the distinction of being one of the last to publish data on associative learning in echinoderms using the Pacific starfish, *Pisaster giganteus*. Associations between light and food (an injured mussel, *Mytilus*) were established. Normally, starfish remained on the walls of laboratory tanks near the water line and when the injured mussel or other food was placed on the tank floor, they went to the bottom to obtain it. Trials consisted of removing the top of the tank (the animals had been housed in the dark), turning on a light, and simultaneously placing the injured mussel at the bottom of the tank. A trial lasted for 15 min, and the number of responses occurring during this period were recorded. The first 8 trials consisted of the simultaneous presentation of light and food, trials 9–18 were test trials in which the mussel was placed in the tank after the starfish had responded, and in trials 19–30 the light was turned on and mussels were not presented until 12 hr after the light had been turned off. The latter were tests to determine if experimental extinction could be obtained. The results are presented in Fig. 8 and demonstrate both the acquisition and the extinction of the response. Controls indicated that *P. giganteus* never moved to the bottom of the tank when presented light alone.

Landenberger also discussed whether *P. giganteus* might use this associative ability and noted that the animals were found under the piers, where mussels that grow on the pilings are likely to fall. No starfish were found 5 ft beyond either side of the pier, and it is suggested that associations between a boundary and food on the substrate would be reasonable in natural conditions.

Corson's (1974) unpublished studies also attempted to establish stimulus associations. In a Y-maze, both food reinforcement (a mussel) and electric shock in combination with light and texture stimuli failed to yield positive results. Additionally experiments that investigated the ability of a starfish to learn a one-way and two-way shuttle response with light as CS and shock as the UCS were inconclusive.

Rockstein and Spritzer (1960) attempted to reverse the daily rhythm of a positive phototaxis in *A. forbesi.* During the daytime the animals were positively phototropic. For five weeks they were exposed to light during the nighttime hours. There was no change in the rhythmic phototaxis, and the daytime responsiveness also disappeared.

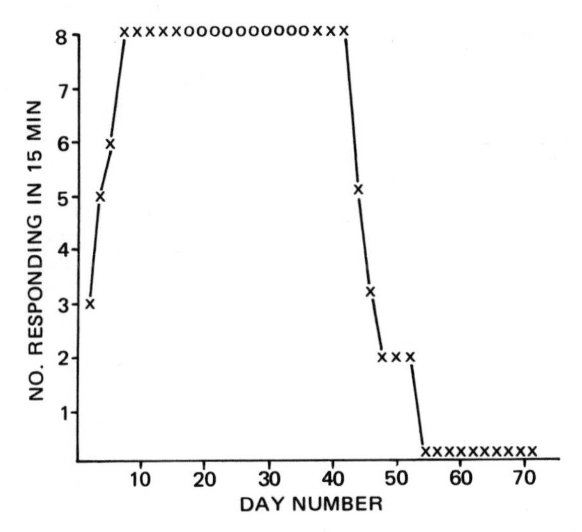

Fig. 8. Association of light and food by the starfish *Pisaster giganteus*. Eight animals were used. Stimulus was a 15-min exposure to light. Exposures were made every 48 hr. On the first eight experimental days, mussels were presented at the beginning of the exposure period (number responding shown by *x*). On the next ten experimental days mussels were given at the end of the exposure period (number responding shown by open circles). After that, mussels were given 12 hr after the exposure period. (Landenberger, 1966, reproduced by permission.)

V. CONCLUSIONS

The phyletic peculiarities and distinctiveness of the echinoderms, and their operations along unusual temporal and behavioral dimensions, have probably led to a notable lack of interest in their learning capacities. Considering the ease with which some learning paradigms might be explored (habituation, for example) at both behavioral and electrophysiological levels, this is surprising. Nerve-muscle preparations have been developed (Cobb and Laverack, 1966; Pentreath and Cobb, 1972) that would be amenable to electrophysiological analyses of simple forms plasticity. Additionally, there is a need for research on larval behavior, in which capacities for adaptation and movement may be more apparent. For example, Dan and Dan (1941) note in *Comanthus japonicus* that larvae which fail to affix themselves during a 3- to 5-day period appear unable to do so later on. It would be interesting to explore the response and habitat preference development in larval forms—it is at this period that "plasticity" might be more essential.

Definitive conclusions concerning echinoderm learning must be forestalled until better preparations and perhaps more patient investigators appear. The existing evidence is certainly suggestive, and with increasing interest in the ecology and nature of marine life, it is hoped that future reviews will contain more extensive data.

REFERENCES

Allee, W. C., 1958, "The Social Life of Animals," Beacon Press, Boston.

Barnes, R. D., 1968, "Invertebrate Zoology," W. B. Saunders, Philadelphia.

Binyon, J., 1972, "The Physiology of Echinoderms," Vol. 49, Pergamon Press, Oxford.

Boolootian, R. A., 1966, "Physiology of Echinodermata," Interscience Publishers, New York.

Bullock, T. H., 1965, Comparative aspects of superficial conduction systems in echinoids and asteroids, *Am. Zool.*, *5*, 545–562.

Campbell, A. C., and Laverack, M. S., 1968, The responses of pedicellariae from *Echinus esculentus*, *J. Exp. Mar. Biol. Ecol.*, *5*, 191–214.

Clark, A. M., 1962, "Starfishes and Their Relations," British Museum (Natural History), London.

Cobb, J. L. S., and Laverack, M. S., 1966, The lantern of *Echinus esculentus*. I. Gross anatomy and physiology, *Proc. R. Soc. Lond. Biol. Sci.*, *164*, 624–640.

Corson, J. A., 1974, Personal communication.

Cowles, R. P., 1910, Stimuli produced by light and by contact with solid walls as factors in the behaviour of ophiuroids, *J. Exp. Zool.*, *9*, 387–416.

Crozier, W. J., 1915, The sensory reactions of *Holothuria surinamerisis*, *Zool. Jahrb. Abt. Allg. Zool. Physiol.*, *35*, 233–297.

Dan, J. C., and Dan, K., 1941, Early development of *Comanthus japonicus*, *Jap. J. Zool.*, *9*, 565–574.

Davenport, D., 1966, "Echinoderms and the control of behavior in associations," *in* "Physiology of Echinodermata" (R. A. Boolootian, ed.), pp. 145–156, Interscience Publishers, New York.

Diebschlag, E., 1938, Ganzheitliches Verhalten und Lernen bei Echinodermen, *Z. Vgl. Physiol., 25,* 612–654.

Fell, H. B., and Pawson, D. L., 1966, General biology of the echinoderms, *in* "Physiology of Echinodermata" (R. A. Boolootian, ed.), pp. 1–48, Interscience Publishers, New York.

Fraenkel, G. S., and Gunn, D. L., 1961, "The Orientation of Animals," Dover Publications, New York.

Glaser, O. C., 1907, Movement and problem solving in *Ophiura brevispina, J. Exp. Zool., 4,* 203–220.

Hyman, L. H., 1955, "The Invertebrates," Vol. 4, McGraw-Hill, New York.

Jennings, H. S., 1907, Behaviour of starfish *Asterias forreri, Univ. Calif. Publ. Zool., 4,* 53–185.

Kjerschow-Agersborg, H. P., 1918, Bilateral tendencies and habits in the twenty-rayed starfish, *Pycnopodia helianthoides. Biol. Bull. (Woods Hole), 35,* 232–253.

Landenberger, D. E., 1966, Learning in the Pacific starfish. *Pisaster giganteus, Anim. Behav., 14,* 414–418.

Maier, N. R. F., and Schneirla, T. C., 1935, "Principles of Animal Psychology," McGraw-Hill, New York.

Mauzy, K. P., Birkeland, C., and Dayton, P. K., 1968, Feeding behavior of asteroids and escape response of their prey in the Puget Sound region, *Ecology, 49,* 603–619.

Moore, A. R., 1910, On the righting movements of the starfish, *Biol. Bull. (Woods Hole), 19,* 235–239.

Moore, A. R., 1921, Stereotropic orientation of the tube feet of starfish (*Asterias*) and its inhibition by light, *J. Gen. Physiol., 4,* 163–169.

Nichols, D., 1962, "*Echinoderms,*" Hutchinson, London.

Ohshima, H., 1940, The righting movements of the sea-star *Oreaster nodosus* (Linne), *Jap. J. Zool., 8,* 575–589.

Olmsted, J., 1917, The comparative physiology of *Synapta hydriformis* (Lesueur), *J. Exp. Zool., 24,* 333–379.

Pentreath, V. W., and Cobb, J. L. S., 1972, Neurobiology of echinodermata, *Biol. Rev. (Camb.), 47,* 363–392.

Preyer, W., 1887, Uber die Bewegungen der Seesterne, *Mitt. Zool. Stat. Neapel., 7,* 27–127.

Reese, E. S., 1966, The complex behaviour of echinoderms, *in* "Physiology of Echinodermata" (R. A. Boolootian, ed.), pp. 157–218, Interscience Publishers, New York.

Rockstein, M., and Spritzer, R., 1960, Light orientation in the starfish, *Asterias forbesi, Anat Rec., 108,* 379 (abstract).

Smith, J. E., 1965, Echinodermata, *in* "Structure and Function in the Nervous Systems of Invertebrates" (T. H. Bullock and G. A. Horridge, eds.) Vol. 2, pp. 1519–1558, W. H. Freeman, San Francisco.

Sokolov, V. A., 1961, Tactile conditioning in the starfish *Asterias rubens* (in Russian), *Murmanskii morskoi Biologicheskii Institut. Trudy, 3,* 49–54.

Takahashi, K., 1964, Electrical responses to light stimuli in the isolated radial nerve of the sea urchin; *Diadema setosum* (Leske), *Nature* (*Lond.*), *201,* 1343–1394.

Uexkull, J. V., 1897, Vergleichend sinnesphysiologie Untersuchungen II. Der Schatten als Reiz für *Centrostephanus longispinus, Z. Biol., 34,* 319–339.

Ven, C. D., 1921, Sur la formation d'habitudes chez les astérias. *Arch. Néerl. Physiol., 6,* 163–178.

Warden, C. J., Jenkins, T. N., and Warner, L. H., 1940, "Comparative Psychology," Vol. 2, Ronald Press, New York.

EDITOR'S NOTE

In the interest of encouraging discussion and objective debate over the demonstrations of invertebrate learning reviewed in these volumes, we include at this point a commentary by M. E. Bitterman. Professor Bitterman has had a long involvement with issues of animal learning and his significant contributions to comparative psychology have had considerable impact on research directions. Specifically, we were interested in obtaining a critical examination of the invertebrate learning literature in light of the assumptions, techniques, and findings emerging from vertebrate research. Bitterman's high standards and impatience with the continued publication of inadequately run studies have led to a conservative and pessimistic view, a perspective that is partially justified on the basis of past errors and technical limitations. His major points are that (1) we know better than to run studies purporting to demonstrate learning without incorporating adequate controls and (2) the use of techniques to rule out experimenter bias should now be a common occurrence.

While we share Bitterman's views regarding the importance of maintaining a cautious and skeptical attitude in evaluating the research literature on invertebrate learning, such skepticism can be carried to such extremes as to be counterproductive for advancing the scientific enterprise. It is reminiscent of the archetypal skeptic who when informed that "there is a black cow in the field," replies, "Well it appears to be black on this side." In questioning the validity of the demonstration of the partial reinforcement effect in earthworms, Bitterman fails to recognize a more plausible, if less contentious, alternative interpretation of the significant first trial decrement of the CR group in the Peeke *et al.* experiment. For example, if the phenomenon were taken at face value, rather than inferring experimenter bias it would be reasonable to think of it as reflecting differential retention of the effects of reinforcement. This interpretation would be entirely consistent with the Ratner-Miller experiments on the effects of intertrial interval on retention (Vol. 1), since the PR group would have longer interreinforcement intervals than the CR group. Furthermore, it should be noted that Dyal has previously called attention to the importance of a trial-by-trial presentation of data in the evaluation of retention processes in the very experiments that Bitterman is criticizing. On p. 265 of the annelid chapter, he says, "It is unfortunate that these data were not analyzed to determine the relative amount retained in the first few trials during a daily session. It would not be surprising to find that there was very substantial forgetting between training sessions as judged by performance on the first one or two trials since this sort of effect has been obtained both in habituation in polychaetes (Dyal and Hetherington, 1968) and in instrumental conditioning of earthworms (Datta, 1962)."

There are good examples of where much of Bitterman's challenge has been answered. Examples of objective and automated data collection can be seen in Harden's habituation work with *Stentor,* in which photographic records were obtained and in which the most stringent controls were incorporated (Vol. 1, p. 67), and in Lee's operant conditioning of *Aplysia* (Vol. 2, p. 240). There are many others to be found in each chapter, and while there are also many examples of poorly run and controversial studies, our purpose was to communicate "the state of the art" and our optimism reflects our desire to see more research achieve the standards that should be commonplace. It is our considered opinion that Professor Bitterman's challenge can be met and that there is much in the literature from which we can learn.

W. C. Corning
J. A. Dyal
A. O. D. Willows

Chapter 13

CRITICAL COMMENTARY

M. E. BITTERMAN

Laboratory of Sensory Sciences
University of Hawaii
Honolulu, Hawaii

About 35 years ago, when I was an undergraduate student in Schneirla's laboratory and just beginning to be interested in problems of learning, Warden *et al.* (1940) published the second volume of their trilogy on comparative psychology—the volume on *Plants and Invertebrates*—which summarized what we knew then about invertebrate learning. Clearly, we did not know very much, and it must be admitted that we do not know much more today. Habituation is found in a wide range of invertebrates, but still we are in the dark about its relation to associative learning, glimmerings of which also appear almost everywhere, although the controls usually are so inadequate and the measures so subjective that we are hard put to decide what the results mean. Only with few invertebrates have we progressed beyond the turn-of-the-century question as to whether they learn at all. A notable exception, of course, is the octopus, which has been studied in a variety of complex experiments patterned after those with vertebrates, but by a technique so unreliable on the whole (Bitterman, 1966; Walker *et al.*, 1970) that we cannot have any confidence in the findings.

Our lack of knowledge is due not so much to lack of effort as to lack of sophistication on the part of those who have worked with invertebrates. Consider, for example, the large number of classical conditioning experiments without proper controls for the effects of exposure to CS and US apart from their temporal contiguity. To ask for such controls is not (as sometimes has been charged) to deny that sensitization is an interesting phenomenon in its own right, but merely to suggest that the effects of pairing should be distinguished from the effects of stimulation per se. Awareness of the importance of this distinction frequently is evident in the published reports, the lack of proper control procedures reflecting only the inability of the authors to cope with the problem. I remember very well the mingled feelings of impatience and amusement with which I read the paper by Gelber (1952) that provoked "the great paramecium controversy." Did the tendency of the animals to cluster about a sterile wire increase after experience with a bacteria-laden wire simply because the nutritive level of the animals was changed by ingestion of the bacteria? No, Gelber argued (p. 61), since only about 1500 bacteria were added in the course of training to a medium already containing about 180,000 bacteria. Why, then, should the addition of a mere 1500 bacteria have any appreciable "reinforcing effect"? That, said Gelber, was because the 180,000 bacteria already present were on the cilia of the subjects "where they would not be available for food." (Round and round she goes and where she stops nobody knows.) Later, without a trace of amusement but only impatience, I read the paper by Thompson and McConnell (1955) that provoked The Great Planarian Controversy. The authors, who had the benefit of an excellent course in learning (if I do say so myself), should have known better—as should the editor of the journal in which both their paper and Gelber's were published. Is it really so difficult to understand that CS-alone and US-alone groups do not, even together, control for the effects of stimulation per se on an experimental group exposed to both stimuli? I must respectfully disagree with Corning and Kelly, who assert in Vol. 1 that the planarian literature "lays bare the problems inherent in psychological learning theory" (p. 217). What that literature (along with the entire literature of invertebrate learning) lays bare, I should say, are the shortcomings of the work it describes.

Not only are the controls employed in the search for invertebrate learning generally inadequate, but the techniques are crude and subjective, stimuli being presented manually and responses detected visually by observers. Rarely do we encounter even a partially automated experiment or any appreciation at all of the importance of automation. Is it not anachronistic in this electronic age to find Willows complaining in Vol. 2 that experiments were not "done blind to eliminate experimenter bias" (p. 264)? If we chose to consider only reports of experiments either automated or "done blind," we should be dealing with a restricted literature indeed. There is the question, too, of just how blind is blind. In their work on partial reinforcement in the earthworm, for example, Peeke *et al.*, (1965) took "special precautions . . . to ensure that the identity of individual *S*s" in the consistent and partial groups "was unknown to the *E*s during the extinction trials" (p. 567); yet their Fig. 2 (p. 569) reveals a difference between the groups (opposite in direction to the difference between them in the last ten acquisition trials) from the very first trial of extinction—that is, the consistent group showed a marked decrement in responding before having any experience at all with nonreinforcement. It is easy for me to think that the *E*s were not quite blind and that their observations were influenced by anticipation of a conventional partial reinforcement effect.

In some quarters, there is a substantial aversion to automation—witness Thorpe congratulating Ratner "on the fact that he had watched his animals under natural conditions" (since Ratner's training situation for the earthworm consisted of a piece of clear plastic tubing encircling a buzzer and periodically illuminated by a flood lamp, with the experimenter himself looming large on the horizon, it must be evident that naturalness, like beauty, is in the eye of the beholder) and expressing the hope that "he would not automate his recordings" (Thorpe and Davenport, 1965, p. 108). It is interesting to ask what the results of Peeke and colleagues (who used Ratner's technique) would have been had they used an objective indicator of response. In any case, automation does not obviate watching, as I have noted elsewhere (Bitterman, 1965); having given over the responsibility for programming an experiment and taking routine data to equipment far more capable for such purposes than he, the experimenter is free to watch his animals more intently than ever before. Usually, however, automation is not rejected on principle, but because it is mistakenly thought not to be worth the effort. When some years ago I criticized the technique being used in Naples to study discriminative learning in the octopus, a colleague at the Stazione (the only one who seemed to have any real appreciation of the limitations of the technique) pointed out that he had only a short time to spend there each year and that time devoted to improving the technique would yield no data. While the manual-observational data left much to be desired, he argued, they were better than no data. I could not agree. Automation, I insisted, would not only yield data worth collecting—which his, I thought, were not—but would vastly increase the efficiency of the data collection processes, maximizing the productivity of his short stays in the laboratory.

Although the summaries of the literature provided in these volumes, even if taken at face value, give us little reason to brag about our knowledge of invertebrate learning, they in fact present a more optimistic picture than the papers summarized seem to me to warrant. In checking the sources of confident statements made in the text, a critical reader may expect to be disappointed repeatedly. For example, Dyal tells us in Vol. 1 that the partial reinforcement effect has been demonstrated in earthworms, citing a paper earlier than the one to which I have already made reference and which does not present trial-by-trial extinction curves (Wyers *et al.*, 1964), but the trial-by-trial curves published later, which are based in part on the earlier data and which cast doubt on the whole enterprise. Whatever the reason for this slip of Dyal's, such inquiries soon make it quite apparent that a great deal of smoothing over and cleaning up has been accomplished in the summaries, owing perhaps to an understandable tendency on the part of authors enthusiastic about their areas of research to accentuate the positive. The picture becomes even more gloomy, however, when we consider (as we must) that a great deal of smoothing over and cleaning up already has been accomplished in the experimental reports themselves. For example, the critical trial-by-trial extinction curves were not presented in the first of the two cited papers on partial reinforcement in earthworms. (They were presented in the second only at the request of an editorial consultant, although their significance even then was lost upon authors and editor alike.) The training procedures described in published reports also seem less objectionable very often than they are in reality, as I can testify from visits to laboratories in which the research is done.

It was tremendously instructive for me to observe firsthand the Naples octopus work which I had judged on the basis of published reports to hold such great promise. The method as described in the literature had seemed so reasonable—the octopus waiting in its "home" at the back of the tank as each trial began, a stimulus on a rod introduced at the front of the tank by the experimenter, the animal attacking the stimulus, the latency of attack being measured with a stopwatch, and the response rewarded with food or punished with shock. The published reports had not prepared me for the fact that the animal often refused to wait quietly at home, advancing to the front of the tank before the introduction of the stimulus and having to be driven back by the experimenter, who would beat on the tank with a clenched fish or some heavy object. Nor had I been prepared for the difficulties involved in the definition of response. If a full-fledged attack is permitted, the experimenter often can recover the stimulus only by the exertion of considerable force (Dilly *et al.*, 1964) in a contest of wills traumatic both for him and for the animal. The alternative is to reinforce the animal for behavior short of attack, but then, of course, the judgment becomes subjective indeed. Only later did I discover just how aversive the laboratory situation was to the animals. At the time, I knew only that great pains had to be taken to prevent them from escaping, which often they managed to do anyway, the official explanation being that octopuses are "curious" and "like to go exploring." The photograph of two hapless escapees shown in Fig. 1, which is taken from Dilly *et al.* (1964), epitomizes for me the state of things in Naples as I found them in 1960. Of course, there may have been considerable improvement in technique since 1965, when I last worked there. I myself am responsible for at least one change, my substitution of latches for the building bricks formerly used to weight the covers of the living tanks being generously applauded as a striking piece

Fig. 1. The caption for this photograph, which is taken from Dilley *et al.* (1964, p. 92), reads as follows: "The two octopuses that escaped from a tank ($63 \times 53 \times 55$ cm) with a fitted lid (5.2 kg) and four building bricks (3.65 kg. each), the total weight being 19.8 kg. The weight of the two octopuses was 0.48 and 0.35 kg. They could not release their mantles and died."

of American ingenuity. There is, however, not much evidence of improvement in the literature (which is understandable, I suppose, since the deficiencies were not widely recognized in the first place), and I cannot tell at what point, if any, we may begin to have increased confidence in the results.

Why do we study invertebrate learning? Two main interests are apparent in the literature. One is in the evolution of learning, with particular reference to the problem of convergence, comparison of vertebrates and invertebrates affording an opportunity to examine independently developed mnemonic solutions to common problems of adjustment. Certain gross resemblances are perhaps to be expected; given the laws of physics and the properties of available building materials, there are, as Pantin (1951) has noted, only so many ways to build a bridge. Identities would be rather surprising, however, and I can only wonder at the readiness of Wells in Vol. 2 to conclude that "bees do not appear to show . . . learning phenomena qualitatively different from those exhibited by rats" (p. 183) or of Sanders in the present volume to suggest that the study of octopuses in experiments patterned after those done with rats "has aided the development of a general theory of animal discrimination learning" (p. 92). The theory to which Sanders refers has, in fact, scarcely any credibility even with respect to the data for laboratory rats that suggested it (Bitterman, 1972). The other main interest apparent in the literature (clearly linked to the evolutionary interest) is in the mechanisms of learning, which it seems reasonable to examine first in relatively simple systems. Both interests must lead inevitably to systematic functional analysis, for which the development of efficient and objective behavioral methods will be required. In preparation for this task, some greater familiarity with the vertebrate learning tradition than now is evident on the part of those working in the field would be invaluable. Unfortunately, however, we are seeing instead a diminishing respect for that tradition among comparative psychologists, who find it fashionable now to denigrate their past (about which they know less and less) and to stand in awe before the loose anthropomorphic and teleological models of the ethologists. Quite another direction will have to be taken if progress is to be greater in the next 35 years than in the last.

REFERENCES

Bitterman, M. E., 1965, Phyletic differences in learning, *Amer. Psychol. 20*, 396–410.

Bitterman, M. E., 1966, Learning in the lower animals, *Amer. Psychol. 21*, 1073.

Bitterman, M. E., 1972, Review of *Mechanisms of Animal Discrimination Learning* by N. S. Sutherland and N. J. Mackintosh, *Amer. J. Psychol., 85*, 301–303.

Dilly, N., Nixon, M., and Packard, A., 1964, Forces exerted by *Octopus vulgaris*, *Pubbl. Stn. Zool. Napoli, 34*, 86–97.

Gelber, B., 1952, Investigations of the behavior of *Paramecium aurelia:* I. Modification of behavior after training with reinforcement, *J. Comp. Physiol. Psychol., 45*, 58–65.

Pantin, C. F. A., 1951, Organic design, *Adv. Sci., 8*, 138–150.

Peeke, H. V. S., Herz, M. J., and Wyers, E. J., 1965, Amount of training, intermittent reinforcement and resistance to extinction of the conditioned withdrawal response in the earthworm *(Lumbricus terrestris)*, *Anim. Behav., 13*, 566–570.

Thompson, R., and McConnell, J. V., 1955, Classical conditioning in the planarian, *J. Comp. Physiol. Psychol., 48,* 65–68.

Thorpe, W. H., and Davenport, D. (eds.), 1965, Learning and associated phenomena in invertebrates, *Anim. Behav., 13,* Suppl. 1, 1–190.

Walker, J. J., Longo, N., and Bitterman, M. E., 1970, The octopus in the laboratory: Handling, maintenance, training, *Behav. Res. Methods Instrum., 2,* 15–18.

Warden, J. C., Jenkins, T. N., and Warner, L. H., 1940, "Comparative Psychology," Vol. 1, "Plants and Invertebrates," Ronald Press, New York.

Wyers, E. J., Peeke, H. V. S., and Herz, M. J., 1964, Partial reinforcement and resistance to extinction in the earthworm, *J. Comp. Physiol. Psychol., 57,* 113–116.

Chapter 14

SYNTHESIS: A COMPARATIVE LOOK AT VERTEBRATES[1]

R. LAHUE AND W. C. CORNING

Department of Psychology
University of Waterloo
Waterloo, Ontario

The comparative researcher's frame of reference usually encompasses more than an interest in an animal for its own sake. We hope to arrive at some general principles that will be applicable to other systems, and in psychology this "other system" is usually man. Claims that interest in invertebrates can be pursued without a vertebrate perspective are difficult to support when the research is organized around vertebrate paradigms and is concerned with vertebrate categories such as "learning," "aggression," "social behavior," etc. With this anthropocentric bias it is easy to understand why the invertebrates remain the "forgotten majority;" yet, in spite of the Procrustean beds that we force them into, invertebrates appear to have achieved some vertebratelike learning capacities to a remarkable degree.

As documented in the present volumes, many of the usual vertebrate-generated learning paradigms can be successfully demonstrated in one invertebrate species or another, and assuming that we are not overly dogmatic about the uniqueness of vertebrate capacities, these demonstrations should serve to stimulate increased usage of invertebrate models, particularly as cockroaches, snails, and planarians are cheaper and smaller and do not require the scrutiny and bureaucratic intervention of animal care committees. However, in spite of the documentation appearing in these

[1] Preparation of this paper was assisted by a grant from the National Research Council of Canada. Portions of this paper were presented by W. Corning at the conference "Evolution of the Nervous System and Behavior," held at Florida State University, 1973, and will subsequently be published by B. H. Winston Co.

volumes, in no case has a single species been as exhaustively analyzed as some of the common mammalian preparations, and at present, caution must be exercised in drawing comparative generalizations about the similarities and dissimilarities of vertebrates and invertebrates. A full range of vertebratelike capacities may be observed within a particular group of invertebrate species, but no one species as yet demonstrates all capacities. Just as critically, we must consider the problem of species selection and the fact that very few of the existing invertebrates have been studied. In many cases, research with particular invertebrate preparations has been pursued because positive demonstrations of vertebratelike capacities seemed likely.

Much of the impetus for invertebrate research stems from a search for model preparations which are simpler than vertebrate preparations but in which vertebratelike capacities can be explored. While complex in itself, the behavior of most invertebrates is "simplified" when compared to the vertebrate at certain points in organism–environment transactions, and these simplifications can be misleading from a comparative point of view. For example, in the invertebrate there is a considerable degree of sensory abstraction and filtering, and the resulting information is processed by fewer control cells. There is less variability (noise) and the ease with which correlations are established between control units and particular functions can lead to conclusions that have little to do with vertebrate brains. Yet, behaviorally, many functional similarities are achieved. What is needed, of course, are behavioral and physiological investigations that are sensitive to the differences in mechanisms, logical capacities, and functions. Similarity at one level of functional analysis is an insufficient basis for a comparative science and it is necessary to determine at how many levels or at how many points in the information-processing route the invertebrate and the vertebrate are similar or dissimilar. As we will demonstrate below, and as the chapters of these volumes demonstrate, at superficial levels the invertebrate has achieved a considerable array of vertebratelike functions.

I. EXAMPLES OF FUNCTIONAL CONVERGENCE

A. Equipotentiality

Lashley's (1950) principles of "equipotentiality" and "mass action" derived from studies which demonstrated that some complex psychological properties such as learning could not be localized in a specific brain area. The decrement in performance produced by a lesion was found to be proportional to the amount of tissue removed, not the locus. Furthermore, ablation-behavior data frequently suggested a recovery of function after the removal of a "locus" for a particular function. Another example of this effect in the mammalian brain was provided by Adametz (1959), who ablated midbrain tegmentum in two to three stages rather than in one and did not find the comatose animal produced by a one-stage lesion (Morruzzi and Magoun, 1949). There are many other examples of an apparent transfer of a function after damage (see John, 1967, for a discussion). Whether this is a transfer or the utilization of the remainder of what Luria called a "constellation" is not known, but there are many invertebrates that demonstrate a similar process. In *Blatta orientalis,* antenna cleaning is carried out by the front legs. If the front legs are broken off, the cleaning response returns eight days later when the roach will stand on three legs and clean its antennae with the fourth (Luco, 1971). This transfer of function may actually represent more of a Luria "constellation." There may be circuit redundancy so that the function is not actually transferred, i.e., the available and relevant circuitry exists further back in the nervous system and what the animal learns is to balance itself again. The neural paths responsible for this equilibrium learning have been studied and electrophysiological facilitation has been observed in certain units after leg removal. Young roaches show the same transfer phenomenon, and when the legs regenerate, the normal front-leg wiping pattern returns. If the legs are then pulled off once more, the "transfer" occurs immediately. There also are examples of this in spiders, notably the observations of Szlep (1952) in *Aranea diadema.* The cobweb of a spider is a complicated structure to effect and the functions of amputated legs can immediately be assumed by other legs.

Acquired responses also demonstrate the spread of information within the central nervous system. In Horridge's 1962 ingenious roach preparation, the animal or a part of the animal (isolated ganglion preparation) was able to learn to keep one leg flexed in order to avoid shock. When other legs were tested after training, there was a savings in learning. The spread of acquired information is also documented in the planarian regeneration experiments (Corning, 1967). A transfer of habituation has also been noted in *Limulus* (Corning and Von Burg, 1968) and protozoans and rotifers (Applewhite and Morowitz, 1967).

B. Dominant Focus

Roy John (1967) has provided a thorough discussion of an interesting state of sustained excitability termed a "dominant focus" by Ukhtomski. Basically, a dominant focus is achieved when a reflex or some portion of the nervous system is hyperexcited. The changed level of activity in an area of the system is sustained for a period of time after the original stimulation ceases and the area is able to summate input, i.e., formerly inconsequential inputs now have access to the area and can interact with other inputs. As a result of this temporary excitatory state, the effects of sensory inputs and presumably the integrative possibilities are changed. For example, Rusinov applied subthreshold anodal current to an area of motor cortex and then presented the animal with a stimulus (light or sound); during polarization and for a period of time after its cessation, the endogenous stimulus would elicit the reflex appropriate to the motor area that was stimulated (for reviews see Adey, 1969; John, 1967; Morrell, 1961; Razran, 1971). Morrell established that there is some specification to this temporary connection between an external stimulus and a motor response. During application of an anodal current to the cortex, a 200-Hz tone was presented 30 times and a 500-Hz tone was presented only once. Prior to polarization, units in the motor cortex did not respond to any auditory input, but during polarization, units began to respond both to 200-Hz signals and to 500-Hz signals. After polarization was terminated, the units only responded to the 200-Hz tone. The "address" of sensory input would seem to depend upon the area undergoing a heightened level of excitability.

The work of Franzisket (1963) on the frog (*Rana esculenta*) appeared to have demonstrated this phenomenon at a behavioral level in a "lower" vertebrate, but subsequent replication attempts (Corning and Lahue, 1971) showed that the reflex enhancement and stimulus "rerouting" were nonspecific.

There are two phenomena studied in simple systems that may be relevant to the dominant focus demonstrations: posttetanic potentiation (PTP) and heterosynaptic facilitation (HSF). The two appear to be similar in that an increase in excitability alters the effect of input, but they represent different neural mechanisms. For PTP, the basic paradigm is to use a monosynaptic pathway of the cat spinal cord. A test stimulus is delivered to the dorsal root once every 2 sec. This results in a ventral root response of a particular amplitude. Then the dorsal nerve is stimulated at a high frequency for a period of time, after which the test stimulus is again applied. The response of the motor neuron pool is now enhanced and this potentiation can last for several hours depending upon the system and the particular stimulus parameters. Although PTP was initially demonstrated in mammalian spinal cord preparations, similar demonstrations have been reported in the invertebrate (Bruner and Kennedy, 1970).

The phenomenon of heterosynaptic facilitation (HSF) has been extensively explored in a giant cell of *Aplysia* (Kandel and Tauc, 1965). As reviewed by Willows (Chapter 10), the procedure is to first stimulate one pathway and record the EPSP (excitatory postsynaptic potential) in the cell.

Stimulation of another pathway after this initial "priming" stimulation, was observed to increase the EPSP produced by the test stimulus by as much as 700% and for as long as 42 hr (Haigler and Von Baumgarten, 1972).

The consensus is that both PTP and HSF involve alterations in presynaptic events but that the two processes show many dissimilarities with respect to mechanisms. According to Kandel *et al.* (1970), PTP is shorter lasting, sensitive to temperature or the substitution of Li^+ for Na^+, and involves a single pathway (potentiation of one pathway does not "transfer" to others). In HSF, the enhancement is longer, is unaffected by cooling or the substitution of Li^+ for Na^+ and involves multiple pathways. One current speculation is that the priming stimulus increases the activity in a third neuron that terminates upon the presynaptic terminals of the test pathway.

Both PTP and HSF provide mechanisms for short-term "memory." Interpretation concerning the altered stimulus consequence following excitation should be based upon knowledge of the circuitry (i.e., single or multiple pathways), the synaptic physiology, and the actual input pathways that are being activated by the testing and priming stimuli. Since PTP is an increase of synaptic transmission in a single channel, it may represent the basis of nonassociative behavioral changes (sensitization and pseudoconditioning). HSF and the dominant focus phenomenon may be closer to being true associative conditioning mechanisms since the response potentiation is produced by the pairing of stimuli in two different channels. Kristan (1971) has demonstrated in *Aplysia* that pairing the test stimulus with the priming stimulus produced changes that are different from those elicited by the test or priming stimulus alone. To demonstrate a closer parallel with Morrell's findings, a previously unpaired stimulus should be used to determine the specificity of the association.

C. Incremental and Decremental Processes

"Learning not to respond" (habituation) is frequently referred to as a simple form of plasticity, but recent physiological and behavioral findings suggest it is quite complex, and the habituation criteria of Thompson and Spencer (1966) provide a polythetic basis for comparing species (see pp. 7–9, Vol. I). Critical among these criteria is the requirement that fatigue be ruled out, and the usual method for demonstrating this is the use of a different "dishabituating" stimulus. Ideally, when the organism is showing a response decrement to one stimulus as a result of repetition, a different, previously unapplied stimulus will cause a return of the response level to initial values. This demonstration rules out fatigue and also shows that the response decrement is specific to the stimulus characteristics. For example, it is possible to habituate a visual evoked response in cats to a 10-per-sec flicker, i.e., the amplitude of the evoked potential decreases with the 10-per-sec repetitive input. Alteration of stimulus frequency to 6 per sec elicits a full-amplitude evoked response again (John and Killam, 1960).

Pumphrey and Rawdon-Smith (1939) have provided a most unusual example of the specificity of habituation that is possible in an arthropod. Repeated stimulation of the cercal nerve of the roach at a particular frequency resulted in the diminution of the postsynaptic response of a monosynaptic arc; the presynaptic input remained constant. The interpolation of a single extra pulse in the stimulus train of 40 pulses per sec resulted in a sudden increase in the postsynaptic response.

Habituation appears to involve more than one process. Groves and Thompson (1970) have provided an interesting review of the literature which would suggest that two independent and opposing processes are operative during habituation. One process, habituation, involves a decrement in activity of the S-R pathway while the opposing process, central state, sensitizes the response. The two processes interact to yield a particular response level. An increase in central state can prevent habituation from taking place; increased stimulus intensity and frequency are ways to accomplish this. Groves and Thompson cite a number of vertebrate preparations in which these two processes are observed. There is good evidence for these two processes of habituation in an arthropod as well as the possibility of a third, modulatory function (Lahue and Corning, 1973*a*). Dorsal nerve efferents in *Limulus* were activated by several different tactile stimulation frequencies (see pp. 20–21, Vol. 2). With the exception of the slowest (1 per 60 sec), habituation occurred at all frequencies. The central state sensitization process was frequency dependent, being maximal at stimulation rates of 2 per sec and 1 per sec.

Ganglion divisions uncovered a possible third process. Previous research had demonstrated the existence of critical intraganglionic contralateral influences (Von Burg and Corning, 1970). These influences could be disrupted by a ganglion "split." The effect of this disruption on the habituation of dorsal nerve efferents was examined at a stimulation frequency that elicited clear incremental and decremental phases. In one group, the segment that first processes the tactile input was divided; in another group the segment giving rise to the dorsal nerve efferent was divided (output split); in a third, both input and output ganglia were divided; and the fourth group comprised intact abdominal ganglia preparations. The results showed that habituation is best and central state sensitization least when an output ganglion is split. The most sensitization and poorest initial habituation is seen with an input split. The intact group and the group with both ganglia split were the same and showed response changes that lay between the other groups. The effect of an output split was assumed to disrupt the number of influences (central state) on the final common path, while an input split appeared to interfere with a third hypothetical process, a sensory modulation function. Dishabituation tests with a different stimulus (a water drop) showed a much higher response in preparations with an input split, providing further evidence for the existence of a sensory modulation function.

Repetitive input, then, excites an S-R pathway, contributes to and may enhance central state influences on efferent pathways, and increases the operation of sensory modulation processes. If central state influences are not sufficiently large, habituation occurs. The complexity of habituation, then, appears to be as great for the invertebrate as for the vertebrate.

D. Memory Transfer

The invertebrate is not free of the more controversial aspects of brain research; in fact, in the case of "memory transfer," it was an invertebrate preparation that stirred up interest in the phenomenon. The work of McConnell and others on the planarian is reviewed in Chapter 4.

Research with vertebrates such as rats and fish has demonstrated that some sort of transfer effect is also possible in more complex systems. Transfer effects have been obtained with whole-brain injection, RNA fractions, and protein fractions, with the routes of injection being intraperitoneal or intracisternal (for a review, see Fjerdingstad, 1971). The failure by some to demonstrate the effect or to show the specificity of the effect (that is, whether it is a specific transfer of acquired information or a more general excitatory effect) has led to considerable skepticism, but a recent analysis of the literature indicates that the transfer effect is real although the interpretation of the effect is unsettled (Dyal, 1971). It would be easy to conclude that because transfer effects are obtained with both the planarian and the rat, the same mechanism is involved but there are little data that clearly show what the mechanisms or the critical carriers for the transfer effect are. In fact, one is hard pressed for data that prove the injected fractions get into the brain cells of the hosts.

E. Unit Conditioning

One of the pioneers of unit conditioning, Olds (1963), has described a procedure by which the spontaneous firing rate of a neuron may be driven up or down depending upon contingent reinforcing stimuli delivered to "positive" or "negative" reinforcing areas in the hypothalamus of cats. Similarly, the operant conditioning of unit activity has been demonstrated by Hoyle in the locust (see pp. 156, Vol. 2). Certain muscles of a leg in the locust are controlled by a single excitatory axon, and accordingly, the activity of muscle units reflects the output of a single neuron. Since the particular muscle (metathoracic anterior adductor) causes the leg flexion, it was assumed that increased muscle discharge frequency indicated leg flexion and a decrease indicated leg extension. A circuit was devised that would deliver a shock to the leg whenever the muscle discharge fell below a certain rate, a paradigm similar to that used by Horridge. With these procedures it was possible to alter the rate of discharge in the muscle.

F. Image-Driven Behavior

The use of "image-driven" behavior to differentiate between species has been reviewed by Beritoff (1971). After animals are shown food in a particular location subsequent behavior indicates that the animal acts as if the food were still there.

Analogous demonstrations have been reported in spiders (see Chapter 6, Vol. 2). When presented with several prey at once, orb-weaving spiders return the prey one at a time to a retreat where they are bound until all the remaining prey are retrieved. Observation of such behavior immediately makes one suspect that the spider remembers not only the presence on the web of an unfinished meal but also its location. Spiders continue to search for prey which have been removed (or escape) from the web. However, since the removal of the fly left a hole in the web which may have been a cue for the spider, Baltzer (1923; also Peters, 1932) carefully removed the fly without disturbing the web or the spider. Nevertheless, after being unable to immediately relocate the prey, spiders searched for as long as 47 min repeatedly returning to the exact site where the prey was initially caught.

G. Unit Specifications

Studies such as those of Lettvin and associates (1959) or Hubel and Weisel (1963), demonstrating the existence of units in frog and cat nervous systems that respond to complex environmental configurations, have stimulated considerable research in the identification and exploration of single-cell functions. Certainly, the ability of a unit to respond maximally to a small, jerky black dot moving into the frog's "striking" range was a startling finding. Such units are also common in the invertebrate. To cite two examples, there are interneurons in the optic peduncle of the rock lobster that respond to various classes of movement, e.g., fast, medium, and jittery movement, and to the movement of light (Wiersma and Yanagisawa, 1971). There are CNS units in crickets that respond to moving targets but not to forced movement of the eye (Palka, 1969).

"Command units," or cells that seem to control complex sequences of reflexes or specific states, are found in invertebrates. In certain cases, single-cell stimulation will occasionally trigger a swimming escape response, even when the stimulation only elicits a single impulse or short burst in the cell (Willows, 1969). Analogous effects are seen in mammalian brains. Sleep, rage, maternal patterns, postural movements, and other behavioral sequences can be triggered by the stimulation of particular sites (Hess, 1957; Fisher, 1956).

H. Biochemical Correlates

The finer biochemical correlates of learning in the vertebrate brain have been the subjects of intensive and frequently controversial research in recent years. Both RNA and protein synthesis seem likely to alter with learning and one popular strategy has employed specific inhibitors of such synthesis to block learning. Both short- and long-term memory in fish can be influenced with such inhibitors as puromycin and cycloheximide (Agranoff, 1967). A related tactic is to analyze brain macromolecules following learning in an attempt to identify specific molecules which may be synthesized due to the experience; Hyden's work on rats is illustrative (Hyden, 1967).

Using the cockroach leg-lifting preparation described previously, workers in Kerkut's lab (Kerkut *et al.*, 1970) have demonstrated the manner in which the two biochemical strategies complement one another in invertebrate studies. At first metabolic inhibitors (cold temperature) and RNA and protein synthesis inhibitors (actinomycin D, congo red, cycloheximide, etc.) were all shown to retard learning. Other experiments showed that the incorporation of labeled uridine into RNA was greatly enhanced in experimental animals. The inhibitors which slowed learning also inhibited the incorporation of labels. Similar effects were demonstrated in the incorporation of labeled leucine into protein. Radioautography and scintillation counting of sequential sections demonstrated that the uptake of labeled uridine was greatest in that portion of the ganglion which contained the interneuronal and motoneuronal connections most immediately involved in flexion of the leg being trained. Drugs known to decrease cholinesterase (ChE) activity facilitated learning.

ChE levels fell during learning but gradually recovered over two to three days (in nondrugged animals), during which period the learned response decayed substantially.

II. DIVERGENT PROPERTIES

The previous list could certainly be greatly extended but truncated as it is, it still demonstrates that the invertebrate has many properties that are usually thought of as being the domain of "higher brain" researchers. However, there are also some differences and these differences are sufficiently dramatic that serious questions must be raised concerning the use of the invertebrate as a "model" for vertebrate functioning—the tendency to equate animals on the basis of performance is readily inhibited when one looks beyond to the underlying structure and physiology. We have already mentioned the tremendous convergence of function that is seen in evolution, yet the neural mechanisms and system organizations of the invertebrates differ from those of the vertebrate in a number of ways. Those differences stressed by various writers are discussed below (Bullock and Horridge, 1965; Cohen, 1967, 1970; Horridge, 1968; Kennedy *et al.*, 1969; Vowles, 1961).

A. Cell Morphology

Invertebrate cells are commonly unipolar with the cell body located at the "rind" or periphery of the ganglion and communicating with the integrative portions within the neuropile via a very small-diameter and usually nonconducting neuronal segment. Their location and morphology may remove them from any direct participation in the integrative activity of the neuron. Most of this activity must be assumed to occur in the central neuropile "core," a network of dendrites, axonal endings, and synapses. Thus, in most invertebrate preparations, the somata of the majority of cells examined are structurally and functionally isolated. The functional significance of this is not as yet clear, but it might indicate a removal of genetic apparatus from any influence or feedback from neuropile activities. The cell body "rind" can be removed in the invertebrate without interfering with neuronal conduction and integrative properties (in some cases, up to several months—see Wine, 1973). Cohen (1970) suggests that this leads to more stereotyped patterns since the genome is less likely to have its readout altered as a consequence of experience. This may also account for the difficulty in finding long-term memory in the invertebrate. The molluscs represent some exception to this picture: cell bodies are invaded by electrical activity and long-term retention is observed.

B. Specification of Cell Type, Structure, and Location

Many cells are readily "identifiable" due to a remarkable consistency within some species with respect to the function, morphology, and location of particular units. This property has enabled Willows (Chapter 10), for example, to map units, i.e., determine their location in the brain and list their functional interconnections. Whether this indicates a high degree of genetic preprogramming with little developmental modulation (epigenesis) remains to be determined.

C. Nerve Process Differentiation

In the vertebrate the dendrites are thick ("fleshy"), bushy (complex), and interdigitate (having numerous spines), and contain typical cell organelles (mitochondria, intracellular membranous structures, etc.); axonal–dendritic differentiations can be made. In the invertebrate, the processes are less differentiated; there are fewer identifiable tracts; the processes are long and thin and ramify widely after emerging from the cell body; and they do not appear to have the protrusions and inclusions that characterize the vertebrate cell and do not appear to have the array of specific synaptic types seen in the vertebrate. Processes may be seen making multiple contacts with other processes. Generally, there is a lack of ordering and grouping of the processes beyond the initial sensory processing stages.

D. Feedback Circuitry

In vertebrates, feedback at all stages of information processing and response execution is typical; in the invertebrate this is not a typical functional property. Animals are more easily "triggered" and maintain the sequence longer in the absence of any corrective environmental feedback (*Limulus,* for example, will continue mating behavior immediately after being pulled out of the ocean as long as an egg-laying female is present). However, primitive first-stage feedback operations that serve to enhance novelty detection and sharpen input are known. Proprioceptive feedback seems in most cases to be limited functionally to the modulation of completely preprogrammed reflexes.

E. Glial Presence

Heavy myelination with nodes of Ranvier in the vertebrate provides insulation between conducting portions of cells as well as high speeds of conduction by saltation. The virtual absence of myelination and nodes in invertebrate systems drastically limits conduction speed while providing the possibility of ephaptic interactions. Modest increases in conduction velocity have been achieved in some instances by larger-diameter ("giant") fibers. A biochemical adjunct function of glial cells in vertebrates has yet to be established for the invertebrates although Levi-Montalcini and Chen (1969) have found that, in tissue cultures, the glia appear to have a trophic function in nerve growth. This lack of heavy glial wrapping in the invertebrate may permit a more readily established functional contact between cells, although as mentioned above, the lack of involvement of the genome may restrict these contacts to a "short-term" status in memory. The lack of glial enclosures also permits the possibility of extrasynaptic influences between neurons.

F. The Numbers of Neurons and Their General Organization

The invertebrate possesses fewer cells and in most systems these cells are positioned in ganglia which are often separated. Insects, for example, are estimated to have 10^5 to 10^6 neurons in their central nervous system. However, only a small percentage of these can be concerned with integrative functions since 80% of the cells are involved with the processing of sensory input (Hoyle, 1970). The brain of the crayfish (*Procambarus clarkii*) has a total of 70,000–80,000 neurons. The ridiculous is observed in rotifers, in which a total of 200 cells is found in the brains of some species and 800–900 cells in the entire nervous system. However, this reduction in cell number does not prevent learning (Applewhite and Morowitz, 1967).

G. Integration/Information Processing

Vowles (1961) has considered the question of whether the invertebrate is really a simpler version of the vertebrate and stresses the following: there is a simpler neural organization on the receptor side leading to a reduction of information that is available to the central nervous system; the number of motor neurons controlling each muscle fiber is small and much of the integration of excitatory and inhibitory events occurs in the muscle tissue rather than in the central nervous system; the motor neurons are involved in many different functions and complex patterns are achieved through temporal sequences rather than simultaneous spatial patterns; in the vertebrate the fine control of patterns is carried on by the brain, whereas in the insect the ganglia seem to regulate excitability but are not involved in coordination of patterns, a process which occurs at the periphery.

The preceding morphological and physiological differences suggest that the invertebrate and vertebrate very likely use differing substrates and mechanisms in learning. In a recent consideration of invertebrate characteristics, Cohen (1970) concludes that with respect to plasticity, the invertebrate's capacity to learn is limited by the lack of effect of electrochemical events on the genetic apparatus (or vice versa) and the smaller numbers of cells. In spite of these limitations, the invertebrates, as previously discussed, have been shown to do remarkably well in achieving vertebratelike organism–environment transactions.

III. SOME COMPARATIVE STRATEGIES AND PROBLEMS

The reasons for emphasizing comparative tactics in learning studies have recently been discussed (Corning and Lahue, 1972). Two of these, which were mentioned above, are particularly important for present considerations. The first deals with the question of *structure-function similarities and differentiations*. Since neural systems and other systems undergo structural and functional alterations during evolution, an examination of the functional capacities of systems which differ in morphological characteristics may result in the establishment of correlations between structures and capacities.

As mentioned above, investigations to date have probably not sampled the invertebrate kingdom in a random fashion. In most cases, *a priori* biases have determined which species were examined. While this places limitations on the generalizations that can be validly drawn concerning the invertebrates as a group, there is some justification for approaching the problem in such a biased way. Clearly, the variety of invertebrate types precludes the evaluation of performance of various species in identical tasks with the same or a similar apparatus, and animals may be selected because of the task characteristics rather than for some other reason. Tests must be designed with consideration for species-typical responsivity and there is little of this relevant information available for most invertebrates. Assuming such information were available, it remains unlikely that investigators would test a wide variety of animals or paradigms when in many cases the likely result would be "negative." Furthermore, such a random-sampling exercise may not even be an intelligent one to undertake. The invertebrates comprise the largest proportion of animal types and exhibit such a variety of forms that one would only expect a considerable degree of divergence among them. Attempting to establish detailed characterizations of the invertebrates as a group may be an ill-advised enterprise when the final generalizations may be vapid and heuristically weak.

However, given the possibility that unwarranted generalizations will not be drawn, *a priori* selection of species for investigation can be strongly defended when a structure-function comparative strategy is to be employed. The criteria for species selection, though, must be sensibly constructed. Little will be gained, for example, by comparing six-legged and eight-legged species in a runway task. A sounder structural logic must be applied. Various species are chosen because they vary along some relatively well-defined parameter, such as degree of cephalization or relative size of corpora pedunculata. Interpretation of the results of such experiments with invertebrates should be somewhat simplified because cephalization in this group generally means the movement of ganglia forward, in contrast to the vertebrates, which add to existing rostral ganglia and shift functions forward (Bullock, 1974). To further simplify matters, at least the initial demonstrations could be sought in animals which vary less on other parameters, such as gross morphology and habitat, than they do on neural structure. Either alone, or in conjunction with a structural logic, an evolutionary logic may be sought. Animals are chosen for study because they presumably occupy different points on the "bush" of phylogenetic complexity. Comparisons could span phyla although sounder comparisons would more likely be made along finer "branches."

An example of a combined structure-evolutionary logic can easily be constructed. All of the arachnida (scorpions and true spiders) are terrestrial, predatory animals. Although they exhibit considerable behavioral and morphological divergence as a class, several are less divergent on some variables than others. Thus, the Scorpionidae (scorpions), Mygalomorphae (tarantulas), Lycosidae and Pisauridae (wolf spiders), Thomisidae (crab spiders), and Salticidae (jumping spiders) comprise an interesting series of animals for such a comparative study. None makes use of a web as a prey-catching device and all belong to the same class. Evidence of their evolutionary status suggests that they vary from least to most advanced in the order given. The central nervous systems of the families mentioned also vary from the primitive scorpion with several abdominal ganglia, to the tarantula with one or two abdominal ganglia, to the completely fused ganglia of the remainder. This group of animals, then, seems to provide an excellent opportunity for exploiting the structure-function comparative strategy while following both structural and evolutionary logic. Of course, there are some differences in sensory capabilities, habitat preferences, hunting patterns, etc., which will need to be taken into account when experiments are designed. Nevertheless, the relative homogeneity of the group should easily outweigh such problems.

We should reemphasize that comparative investigations that demonstrate true differences in capacities between different species are difficult to carry out. Again, species differ in their sensory and motor capacities, they respond to different motivational factors, and endogenous factors such as hormonal fluctuations also differ. How can we possibly equate animals in a particular learning paradigm to determine if they are the same or different? The approach taken by Bitterman (1965) seems to be one useful strategy. Instead of trying to make comparisons based upon a numerical measure (trials to criterion, errors, etc.) comparisons were made on the basis of functional relations. For example, it is possible to compare rats and fish in a spatial reversal task; both animals can be trained to enter the right or left side of a two-part chamber if they are reinforced for just one choice. Reversal of the reinforced side tests a particular functional capacity, and fish were found to be inferior to rats. The argument that this difference represents a difference in motivational state or in other uncontrolled factors can be answered. If hunger level were the differentiating factor, then if the level of food deprivation were varied throughout the functional range of the variable, a point should be found at which the fish is equal to or better than the rat. This strategy assesses the presence or absence of a logical operation in a particular species and is not unlike the approach taken by Piaget in humans (Flavell, 1963) although it is considerably more systematic.

Assessments of capacity differences that are related to complexity differences can be carried out through a comparison of simplified and complex versions of the same system (Corning and Lahue, 1972; Lahue and Corning, 1971*a,b*). Certain invertebrates have proven to be excellent preparations for this sort of comparative strategy, especially many arthropods, because ganglia are readily identifiable and surgical separation of system components is accomplished with relative ease. This strategy does not permit any comparisons to be made between different phyletic groupings, but it does permit comparisons between levels of complexity or the degree of neural relationships. The difficulties encountered in attempts to equate the task for different species are obviated since the receptors are all within the same system; the stimulus input, at least, can be assumed to be constant for simplified and complex versions of the same system. (An application of this strategy is reviewed in Vol. 2, pp. 20–24.)

A second comparative tactic which is critical for the present discussion is the *special advantage strategy,* in which the particular characteristics of a preparation make it especially suited for certain research aims. Typically, the preparation offers a naturally restricted neural network or one that can be restricted surgically. This "simple system" approach implies that there is some preconception about what is to be found (Abraham *et al.*, 1972). Thus, the assumption that a cell can encode information makes the giant cells of many molluscan and anthropod preparations attractive. In their excellent discussion of the simple system approach, Abraham *et al.* (1972) underscore some problems and questions that should be seriously considered in the adoption of this strategy. Again, the major difficulty seems to involve the generalizability of the data. If there is a common mechanism of learning in all organisms, there must be differences in the way organisms use this mechanism. Repression and depression of DNA coding sites may indeed be a basis of long-term changes in cellular function (Gaito, 1972), but it is hard to imagine a protozoan, flatworm, and cat using this mechanism in exactly the same way. The mechanism could subserve a change in molecular conformation at the membrane, an increase in metabolism of a synaptic transmitter, a modified neuronal morphology, etc. It is also possible that at certain levels of analysis (synaptic pharmacology, for example), one could find that a flatworm and a cat both rely upon a depleted transmitter supply to effect, say, the habituation of a particular response. However, the *network architecture* that is eventually responsible for behavioral habituation is completely different in flatworms and cats. Evolutionary divergence may involve the biophysical-biochemical and/or morphological realms. Finally, are the changes observed at the simple-system level generalizable to the intact animal? There is a gamble in the selection of a system for special advantages, such as the brain of *Tritonia* (Willows, Chapter 10), in which there are large, identifiable cells that are most accessible for microelectrode recording—the system may not learn. "Do we pick a behavior capable of exhibiting learning and hope the cellular analysis can follow, or do we optimize a preparation for cellular analyses and hope for the learning to hatch?" (Abraham *et al.*, 1972).

Consideration of these problems in our own invertebrate research has led us to conclude that the best approach is to define and explore behavioral plasticity on the basis of multiple criteria and "characters" and that analysis must include data from several levels of analysis (see pp. 5–6, Vol. 1). This "polythetic" strategy would prevent spurious conclusions and generalizations about similarities in function that derive from "monothetic" analysis. This strategy, although derived from taxonomists, is currently being applied in the behavioral sciences (Pilowsky *et al.*, 1969; Strauss, 1974; Strauss *et al.*, 1973), and there is a concerted effort to use the technique in generating a new psychological "taxonomy" (Corning *et al.*, 1974). When one is making a conclusion about the similarity or common grouping of two different species with respect to some process, the

Table I. A Comparison of Habituation Criteria and Mechanisms

		Criteria[a]	Possible mechanisms
PROTOZOA			
Spirostomum ambiguum	1	Wawrzynczyk, 1937; Kinastowski, 1963*a*,*b*; Applewhite and Morowitz, 1966, 1967	Rate of habituation is independent of temperature, protein synthesis, and RNA synthesis. A passive intracellular diffusion of ions is thought to result in an uncoupling of contractile elements. Recovery is temperature and metabolically dependent, probably involving active ion transport. (Reviewed in Corning and Von Burg, Chap. 2).
	2	Wawrzynczyk, 1937; Kinastowski, 1963*a*; Applewhite, 1968	
	3	Kinastowski, 1963*a*	
	4	Kinastowski, 1963*a*	
	5	Kinastowski, 1963*a*	
	8B	Applewhite *et al.*, 1969; a more intense or a different stimulus applied after habituation training elicited a response	
		Wawrzynczyk, 1937; recovery of response was obtained following a reduction in stimulus frequency	
Stentor coeruleus	1	Wood, 1970*a*,*b*; Harden, 1973	Electrophysiological recordings indicate that habituation is due to a decrement in a sensory receptor function (Wood, 1971).
	2	Harden, 1973; Wood, 1970*a*,*b*	
	3	Harden, 1973; Wood, 1970*a*,*b*	
	5	Wood, 1970*a*,*b*	
	7	Wood, 1972; spatial generalization to the test stimulus which is graded with distance from the habituated site and dependent upon the initial level of response decrement	
	8A	Harden, 1973; stimulation due to changing the water in the test bowl resulted in an enhancement of the habituated response	
	8B	Harden, 1973; a different stimulus (strobe light) resulted in a response following habituation	
		Wood, 1970*a*,*b*; following habituation a different or more intense stimulus elicited a response; however subsequent presentation of the original stimulus did not elicit a full response	
COELENTERATA			
Hydra	1	Rushforth *et al.*, 1963; Rushforth, 1965, 1967, 1973	Stimuli drive a single-event–generating pacemaker which inhibits a
	2	Rushforth, 1967	

Table I. (Continued)

	Criteria[a]	Possible mechanisms
	3 Rushforth, 1965, 1967 4 Rushforth, 1965, 1967 5 Rushforth, 1965, 1967 7 Tests failed to demonstrate any generalization; Rushforth, 1973 8B Rushforth *et al.*, 1963; application of a different stimulus resulted in a response	spontaneously active multiple-event generator; successive stimuli gradually fail to excite the single-event generator.
ANNELIDA Polychaeta *Nereis pelagica*	1 Clark, 1960*a,b*; Evans, 1969 2 Clark, 1960*a,b* 3 Clark, 1960*a* 4 Clark, 1960*a* 5 Clark, 1960*a*; Evans, 1969 7 Clark, 1960*a*; no generalization across modalities 8B Clark, 1960*a*; lack of cross-modal generalization implies this type of dishabituation	Synaptic failures occur at the sensory-giant fiber and giant-motor junctions (Horridge 1959). Decerebration has no effect.
Oligochaeta *Lumbricus terrestris*	1 Kuenzer, 1958; Gardner, 1968 2 Kuenzer, 1958; Gardner, 1968 3 Kuenzer, 1958 5 Kuenzer, 1958 7 Kuenzer, 1958; spatial generalization but no cross-modal 8 Kuenzer, 1958; lack of cross-modal generalization implies this type of habituation	Habituation to tactile stimuli localized at internuncial sensory-giant and giant-motor junctions (Roberts 1962). Decerebration lowers both the threshold to stimulation and the rate of habituation.
Hirudinea *Macrobdella decora*	1 Ratner, 1972 2 Ratner, 1972 3 Ratner, 1972 4 Ratner, 1972	
ARTHROPODA Chelicerata *Limulus polyphemus*	1 Lahue and Corning, 1971*a*, 1973*a*; Corning and Lahue, 1972 2 Lahue and Corning, 1971*a*, 1973*a*; Corning and Lahue, 1972 3 Lahue and Corning, 1971*a*, 1973*a* 4 Lahue and Corning, 1973*a* 6 Lahue and Corning, 1971*a*	An interplay of at least three functional systems which respond to the stimulus result in either habituation, sensitization, or both dependent upon experimental pa-

Table I. (Continued)

		Criteria[a]	Possible mechanisms
	7	Lahue and Corning, 1973*b*; spatial generalization	rameters. Change in temperature drastically affects rate of habituation (Lahue 1974).
	8A	Lahue and Corning, 1973*b*; following a different stimulus the response to the habituating stimulus recovers	
	8B	Lahue and Corning, 1971*a*, 1973*a*; a different stimulus elicits a response following habituation	
	9	Lahue and Corning, 1973*b*	
Crustacea			
Procambarus clarkii	1	Bruner and Kennedy, 1970	In an isolated motor giant-muscle preparation, habituation correlates with a decrease in excitatory junctional potential amplitude; not due to a change in the electrical properties of the postjunctional membrane, inhibitory buildup, or presynaptic transmitter depletion.
	2	Bruner and Kennedy, 1970	
	4	Bruner and Kennedy, 1970	
	8A	Bruner and Kennedy, 1970; interpolation of a short high-frequency stimulus causes the response to the habituating stimulus to recover	
	8B	Bruner and Kennedy, 1970; an increase in stimulus frequency elicits a response following habituation	
Procambarus clarkii	1	Krasne, 1969; Krasne and Woodsmall, 1969	A gradual decrease in the size of the beta component of the EPSP between primary afferents and primary interneuron (Zucker *et al.* 1971); deemed to be presynaptic decrease in transmitter release and is not due to inhibition (Krasne and Roberts 1967); insufficient to explain behavioral habituation (Krasne 1973).
	2	Krasne, 1969; Krasne and Woodsmall, 1969	
	3	Krasne, 1969	
	4	Krasne, 1969	
	5	Krasne, 1969	
	7	Krasne and Woodsmall, 1969	
	8B	Krasne, 1969; a more intense stimulus following habituation elicits a response	
Insecta			
Schistocerca gregaria	1	Horn and Rowell, 1968	Thought to be a depression in excitatory synaptic function due to decreased transmitter release.
	2	Horn and Rowell, 1968	
	3	Horn and Rowell, 1968	
	4	Horn and Rowell, 1968	
	7	Horn and Rowell, 1968; very little demonstrated	

Table I. (Continued)

		Criteria[a]	Possible mechanisms
	8A	Rowell, 1971	
	8B	Horn and Rowell, 1968	
	9	Horn and Rowell, 1968; Rowell, 1971	
MOLLUSCA			
Limnaea stagnalis	1	Holmgren and Frenk, 1961; Cook, 1971	Actively incrementing, hyperpolarizing influences decrease synaptic efficacy (Holmgren and Frenk).
	2	Holmgren and Frenk, 1961; Cook, 1971	
	3	Holmgren and Frenk, 1961	
	4	Cook, 1971	
	6	Holmgren and Frenk, 1961	
	7	Cook, 1971; intramodal generalization	
	8A	Cook, 1971; interpolation of an extra stimulus of a different modality causes a recovery in response to the habituating stimulus	
Aplysia	1	Pinsker *et al.*, 1969; Castellucci *et al.*, 1969	A decrement in EPSP amplitude due to a change in synaptic efficacy resulting from either a decrease in transmitter release or a decrease in postsynaptic sensitivity.
	2	Pinsker *et al.*, 1969; Castellucci *et al.*, 1969	
	3	Carew *et al.*, 1972	
	4	Pinsker *et al.*, 1969; Castellucci *et al.*, 1969	
	5	Pinsker *et al.*, 1969; Castellucci *et al.*, 1969	
	8A	Pinsker *et al.*, 1969; Castellucci *et al.*, 1969; a different stimulus results in a response recovery to the habituating stimulus	
	9	Pinsker *et al.*, 1969; Castellucci *et al.*, 1969	
VERTEBRATA			
	1	Thompson and Spencer, 1966; Wicklegren, 1967*a,b*	Habituation results from the cumulative effects of polysynaptic low-frequency depression (Thompson and Spencer) or a polysynaptic analogue of PTP at inhibitory synapses (Wicklegren).
Cat	2	Thompson and Spencer, 1966; Wicklegren, 1967*a,b*	
	3	Thompson and Spencer, 1966; Wicklegren, 1967*a,b*	
	4	Thompson and Spencer, 1966; Wicklegren, 1967*a,b*	

Table I. (Continued)

	Criteria[a]	Possible Mechanisms
	5 Thompson and Spencer, 1966; Wicklegren, 1967*a*,*b*	
	6 Thompson and Spencer, 1966; Wicklegren, 1967*a*,*b*	
	7 Thompson and Spencer, 1966; Wicklegren, 1967*a*,*b*	
	8A Thompson and Spencer, 1966; Wicklegren, 1967*a*,*b*; a more intense stimulus resulted in an increased response following habituation	
	9 Thompson and Spencer, 1966	

[a] Criteria for habituation as outlined in Thompson and Spencer (1966):

1. Response decrement to repetitive stimulation (habituation).
2. Spontaneous recovery.
3. Potentiation of habituation.
4. Frequency effects.
5. Intensity effects.
6. Below zero habituation.
7. Stimulus generalization.
8. Dishabituation.
9. Habituation of dishabituation.

Note that 8 has been subdivided. 8A refers to a facilitation of a habituated response to the test stimulus following a dishabituation stimulus. 8B refers to a simple response recovery to the dishabituation stimulus alone.

grouping is based upon an aggregate of properties and the more information used, the stronger the comparative statement. According to Sokal (1966) the application of polytheticism in science has been slow to develop because we find it easier to rely upon one or a few characteristics to define a group. Communication is facilitated but accuracy suffers. As an example, consider the findings concerning habituation that are summarized in Table I. Although this table is far from complete, the picture that is emerging is one of considerable divergency. Using behavioral criteria might lead one to the facile conclusion that there is some universality with respect to habituation. This would be true at the behavioral level, but the conclusion is of limited value when there are divergent mechanisms even within species of the same class. The ubiquity of behavioral habituation (including that of plants—see Appendix A) makes comparative statements based only upon behavior of little value—it is similar to announcing that two species are the same because both can fly. As Jensen (1967) points out, man could be characterized monothetically as a featherless biped that talks and would accordingly be in the same grouping as a plucked trained parrot.

IV. THE EVOLUTION OF LEARNING STRATEGIES IN INVERTEBRATES AND VERTEBRATES: SOME SPECULATION

At present, there is little evidence that would permit an accurate portrayal of the differences between invertebrates and vertebrates with respect to learning strategies. As we have argued here, monothetic characterizations coupled with functional convergence lead to oversimplifications concerning differences and similarities. A polythetic characterization requires consideration of biochemical, morphological, and physiological factors as well as behavioral data. However, in the interest of providing a final "vertebrate perspective" we offer a modest speculation concerning the means by which these two divergent groups of animals might achieve general functional equivalence at the behavioral level.

It would seem that there are three general strategies by which adaptive end-states might be achieved:

A. Preprogrammed Morphological/Physiological Strategies

This strategy for existence represents evolutionary adaptations that are mostly independent of environmental input. The reliance is upon the genetic program to provide a sufficient array of behavioral possibilities to meet prevalent environmental contingencies. Experience may enhance or depress the use of various system programs that are available, and hence, the concepts of plasticity or learning are not excluded in this category. However, the genetic readout patterns and the program specifications are determined very early and, ontogenetically, remain relatively immutable. For example, habituation ("learning not to respond") may be conceptualized as a physiological process that results in the behavioral modulation of "species-typical" reflexes. At the cellular level, end-product inhibition represents an example of a metabolic process that can decrease activity in biochemical pathways as a consequence of the accumulation of a particular agent.

B. Epigenetic Strategies

Epigenetic strategies represent processes that are dependent upon environmental input (perhaps at critical developmental stages) for their proper elaboration. In this case the genes provide a "dictionary" or code-reference system that can stipulate a number of species-typical mechanisms, but the system relies upon environmental input for normal development of appropriate neural and behavioral patterns. Inappropriate patterns may occur if the environment is drastically changed. There are many examples of the epigenetic process at all stages of the organism–environment transaction: (1) Retinal cells are observed to degenerate if an animal is deprived of light (Rasch *et al.*, 1961). (2) Stimulation can increase the number of dendritic spines in cortical cells (Schapiro and Vukovich, 1970). (3) Deprivation of environmental interaction but maintenance of stimulus input in young cats produces adults that are not perceptually normal (Held, 1965). (4) "Enriched" environments may improve cortical circulation and glial density in rats (Rosenzweig *et al.*, 1972). (5) Brain polysome concentration is affected by stimulation (Appel *et al.*, 1967). (6) The stimulus preferences and subsequent social relationships of an animal may be biased by early experiences at critical periods (Sluckin, 1965). (7) The development of "vertical line" detectors in a cat's brain may be dependent upon perceptual experience with vertical inputs (Hirsch, 1972). It could be argued that all structures and functions are the result of epigenesis since proper genetic readout requires particular cytoplasmic and extracellular environments. Some have tried to distinguish between physiological and psychological influences in the epigenetic process, but since psychological influences must necessarily be transduced into some sort of physiological language, worrying about such distinctions does not appear to be particularly useful.

C. Acquired Relations Between System Components (Suprastructural)

This route to intelligence assumes the existence of a structural substrate but refers to adaptive transactions that result in new arrangements of communication between various parts. The information that is stored about some experience is represented in the new relationship, need not involve any structural modification, and, in fact, may not be dependent from one point in time to the next upon any specifically localized structures. The sum total of contributions from various portions *is* the information and the relative contribution from any one center may vary. This intelligence mechanism can occur regardless of whether the system has depended upon (1) genetic preprogramming or (2) epigenesis during development. E. R. John's (1967, 1972) electrophysiological investigations of learning provide a good example of this relational strategy in mammals. John's data have led to the formulation of a "statistical configuration" theory, which states that "*the common mode of activity of large numbers of neurons in anatomically extensive systems represents information about a learned experience. Different brain regions come to share a common mode during learning*" (John, 1972, p. 863). (Details of these considerations will be provided below.) Processes such as these represent a means by which the integration of activities can generate new abstractions and concepts.

The first two strategies indicate a cellular–connectionistic assumption, and it is more than coincidence that this assumption finds verification in invertebrate preparations in which there are large, identifiable cells and concomitantly stereotyped behavioral patterns. Additionally, there is vertebrate-based evidence in developmental neurobiology demonstrating that a highly organized and precisely attuned nervous system can be observed unfolding. Functionally, this specificity is also apparent; there is something logically satisfying about finding a particular unit in the frog that responds rather clearly to a "fly" (Lettvin *et al.*, 1959) when there is corresponding behavioral evidence that frogs actually respond to flies. On the output side, electrical tickling of specific sites in the mammalian brain can elicit behavioral sequences independent of experience (Fisher, 1956), and the ultimate in control is seen in the invertebrate sea slug *Tritonia,* in which a single pulse popped into an identifiable cell occasionally elicits a rather complex behavioral sequence that looks remarkably like a starfish escape response (Willows, 1969). Mapping cells, anatomically and functionally, is possible and many investigators are discovering that the invertebrates provide some of the better preparations for this strategy.

Conversely, if one assumes that learning is represented in a coherent *pattern of activity* rather than in a particular connection or set of connections, there are evolutionary products (perhaps only mammalian brains) that seem to operate on this principle. The earlier ablation work of Lashley (1950) provided considerable impetus for a modern movement against the cellular–connectionistic doctrine, and the previously mentioned electrophysiological work of John (1967, 1972) continues to produce a strong challenge. John's objections are essentially based upon the following observations: (1) The results of ablation studies are inconsistent with respect to localizing learning processes. (2) Multiple-lesion studies indicate that some functions such as learning and memory are not eliminated after the destruction of particular brain areas and that therefore the brain does not store these functions in a place. (3) The results of ablations performed early in development also support the above findings. (4) Observations of unit discharge patterns show that there is a great amount of variability in firing rate and that cells can respond to a variety of different modalities—the specification of information in a single neuron would seem unlikely. (5) With respect to studies of "simple systems," the plasticities that are demonstrated are relatively brief and such findings may not apply to vertebrates.

As an alternative to a cellular–connectionistic assumption, John proposes a "statistical configuration theory":

> The critical event in learning is envisaged as the establishment of representational systems of large numbers of neurons in different parts of the brain, whose activity has been affected in a coordinated way by the spatio-temporal characteristics of the stimuli present during a learning experience. The coherent pattern of discharge of neurons in these regions spreads to numerous other regions of the brain. Sustained transactions of activity between participating cells permit rapid interaction among all regions affected by the incoming sequence of stimuli as well as the subsequent spread. This initiates the development of a *common mode of activity,* a temporal pattern which is coherent across those various regions and specific for that stimulus complex. As this common mode of activity is sustained, certain changes are presumed to take place in the participating neuronal populations, which are thereby established as a representational system. Whether such changes are alterations of "synaptic efficiency" or not, it is assumed that the critical feature of these changes is to increase the probability of recurrence of that coherent pattern in the network. Certain types of preexisting neuronal *transactions* become more probable, but no new connections are assumed to be formed (John, 1972, pp. 853–854).

Evidence supporting this theory is derived from both macro- and micro-electrode recordings in the brains of cats. Basically, the data demonstrate that during conditioning, average evoked responses in various anatomically dissimilar regions of the brain show increasing similarity in waveshape as conditioning proceeds. Electrode position is less important in the obtaining of this concordance than are the stimulus characteristics. Comparisons of unit discharges and multiple-unit macropotentials demonstrate a greater invariance during learning in the macropotential readout. Further details of these studies have been recently reviewed (John, 1967, 1972). In general, the data indicate that information representing learning can best be expressed in terms of the coherent activity of widely located ensembles rather than in terms of strictly circumscribed loci.

Whether information processing and the integration of information from various inputs can occur via "suprastructural" processes remains to be thoroughly explored in the invertebrate. Certainly, the generally accepted view is that there is little or no macropotential, slow-wave activity in the invertebrate system (Bullock and Horridge, 1965). However, Hoyle (1970) suggests that large spike potentials of up to 30 mV and 5 msec in duration probably represent the activity of synchronously discharging cells. In extracellular recordings of the neuropile, Hoyle also finds slow negative potentials that precede and accompany bursts of muscle potentials. These could represent summated EPSPs and IPSPs (inhibitory postsynaptic potentials) that bias the thresholds of motor cells. This type of slow potential involvement in integrative activities may represent the ultimate in invertebrate suprastructural possibilities, but the critical difference between the invertebrate and the vertebrate may be that in the former the process is still anatomically locked, whereas in John's data (1972) the macropotentials can be generated by dissimilar structures.

On the basis of existing evidence, it would seem that the vertebrates and the invertebrates differ in two of the three learning strategies: *epigenesis* and *suprastructural information-processing*. Both the chordate phylum and the invertebrate phyla most likely rely upon genetic programming for the development of adaptive strategies. Evidence suggests that the vertebrate brain, while depending upon "hard-wired" circuitry, would depend more upon experiential input for structural specification. The adaptiveness of this epigenetic strategy is obvious: besides the facilitation of neuronal growth, glial density, circulation, etc., by experiences, it is probable that logical operations can be stipulated by these experiences. The organism is limited by its ancestry with respect to capacity, but selective elaboration of its cellular processes is dependent upon the particular environment it finds itself in. The degree of epigenesis varies among vertebrate groups: in cats, the deprivation of experience with "verticalness" results in a reduction of vertical detector units in the brain (Hirsch, 1972), but in rabbits, experiential deprivation does not have a similar effect (Ranney Mize and Murphy, 1973). The influence of experience upon structural elaboration may occur quite early in development and then decrease rapidly, or it may be more prolonged and dependent upon "critical periods" of development in various systems of the organism. In invertebrates, one suspects that there is considerable reliance upon genetic preprogramming and that epigenesis is restricted to early stages of development. There is a need for "early experience" studies in invertebrates. For example, would the deprivation of rapid-movement stimuli early in development lead to a reduction of units that respond to these events in adult crustaceans? Some examples of possible early experience effects have appeared (Hazlett, 1971; Thorpe, 1939*a, b*).

The epigenetic route to intelligence could still confine the organism to specific circuit arrangements for its adaptive transactions. Freedom from these constraints has probably been achieved in the mammalian brain by *suprastructural information-processing* mechanisms. The macropotential activity, representing participation by ensembles of neurons in common modes of activity, could add remarkable power to the ability of an organism to encode information and integrate it with past and anticipated events. This process requires certain critical numbers of neurons with a sufficient array of potential interconnections (and perhaps with particular anatomical configurations), but beyond the "critical mass," the capacities are totally dependent upon the individual's experience and more independent of the structure. Circuit redundancy may permit the invertebrate to achieve maximal benefits from the "switchboard" strategy, but the critical limitation remains, i.e., the structural constraints. As with the epigenetic strategy, there is a need for research on the information-carrying capacities of slow potentials in invertebrate systems.

In conclusion, we may suggest a rather depressing possibility for those anthropocentric investigators bent on using invertebrates as models for the mammalian brain: the critical means by which the advanced vertebrate brain processes higher-order information and achieves integration of various sources of information may be totally different than those means used by the invertebrate. Perhaps at lower-order information-processing levels or early in development the more connectionistic strategies are used in the vertebrate. It seems clear to us that a polythetic approach would ensure a sufficient body of information so that the connectionistic assumptions that are so easily verified in the invertebrate will not provide a limited and perhaps erroneous picture of the vertebrate brain. But at the very least, the polythetic study of invertebrates should keep the vertebrate researcher humble: if a few thousand cells, arranged in a completely different manner and with a different phylogenetic origin can achieve some of the same behavioral functions and learning capacities as the vertebrate brain, then we may be asking limited or simple-minded questions in "higher brain" research. On the other hand, exploration of invertebrate learning capacities and the mechanisms by which invertebrates achieve these capacities could be pursued because invertebrates exist and are interesting in themselves. Adoption of this view makes a vertebrate comparison relatively unnecessary and, accordingly, would hopefully remove invertebrate research from the extant vertebrate biases.

REFERENCES

Abraham, F. D., Palka, J., Peeke, H. V. S., and Willows, A. O. D., 1972, Model neural systems for the neurobiology of learning, *Behav. Biol.*, *7*, 1–24.

Adametz, J. H., 1959, Rate of recovery of functioning in cats with rostral reticular lesions, *J. Neurosurg.*, *16*, 85–97.

Adey, W. R., 1969, Slow electrical phenomena in the central nervous system, *Neurosci. Res. Program Bull.*, *7*, 75–180.

Agranoff, B. W., 1967, Agents that block memory, *in* "The Neurosciences" (G. D. Quarton, T. Melnechuk, and F. O. Schmitt, eds.),. Rockefeller University Press, New York.

Appel, S. H., Davis, W., and Scott, S., 1967, Brain polysomes: response to environmental stimulation, *Science*, (Wash. D.C.), *157*, 836–838.

Applewhite, P., 1968, Temperature and habituation in a protozoan, *Nature* (London), *219*, 91–92.

Applewhite, P., Gardner, F. T., and Lapan, E., 1969, Physiology of habituation learning in a protozoan, *Trans. N.Y. Acad. Sci.*, *31*, 842–849.

Applewhite, P., and Morowitz, H. J., 1966, The micrometazoa as model systems for studying the physiology of memory, *Yale J. Biol. Med.*, *39*, 90–105.

Applewhite, P., and Morowitz, H. J., 1967, Memory and the microinvertebrates, *in*, "Chemistry of Learning" (W. C. Corning and S. C. Ratner, eds.), Plenum Press, New York.

Baltzer, F., 1923, Beitrage zur Sinnesphysiologie and Psychologie der Webespinnen, *Mitt. Naturforsch. Ges. Bern.* 163–187.

Beritoff, J. S., 1971, "Vertebrate Memory: Characteristics and Origin," Plenum Press, New York.

Bitterman, M. E., 1965, Phyletic differences in learning, *Am. Psychol.*, *15*, 709–712.

Bruner, J., and Kennedy, D., 1970, Habituation: Occurrence at a neuromuscular junction, *Science*, (Wash. D.C.), *169*, 92–94.

Bullock, T. H., 1974, Comparisons between invertebrates and vertebrates in nervous organization, *in* "The Neurosciences Third Study Program," MIT Press, Cambridge, Mass.

Bullock, T. H., and Horridge, G. A., 1965, "Structure and Function in the Nervous Systems of Invertebrates," Freeman, San Francisco.

Carew, J. J., Pinsker, H. M., and Kandel, E. R., 1972, Long-term habituation of a defensive withdrawal reflex in *Aplysia*, *Science* (Wash. D.C.), *175*, 451–454.

Castellucci, V., Pinsker, H., Kupfermann, I., and Kandel, E., 1969, Neuronal mechanisms of habituation and dishabituation of the gill-withdrawal reflex in *Aplysia*, *Science* (Wash. D.C.), *167*, 1745–1748.

Clark, R. B., 1960*a*, Habituation of the polychaete *Nereis* to sudden stimuli. I. General properties of the habituation process, *Anim. Behav.*, *8*, 82–91.

Clark, R. B., 1960b, Habituation of the polychaete *Nereis* to sudden stimuli. II. The biological significance of habituation, *Anim. Behav.*, *8*, 92–103.

Cohen, M. J., 1967, Some cellular correlates of behavior controlled by an insect ganglion, *in* "Chemistry of Learning" (W. C. Corning and S. C. Ratner, eds.), Plenum Press, New York.

Cohen, M. J., 1970, A comparison of invertebrate and vertebrate central neurons, *in* "The Neurosciences, Second Study Program," Rockefeller University Press, New York.

Cook, A., 1971, Habituation in a freshwater snail (*Limnaea stagnalis*), *Anim. Behav.*, *19*, 463–574.

Corning, W. C., 1967, Regeneration and retention of acquired information, *in* "Chemistry of Learning" (W. C. Corning and S. C. Ratner, eds.), Plenum Press, New York.

Corning, W. C., and Lahue, R., 1971, Reflex "training" in frogs, *Psychonomic Sci. Sect. Anim. Physiol. Psychol.*, *23*, 119–120.

Corning, W. C., and Lahue, R., 1972, Invertebrate strategies in comparative learning studies, *Am. Zool.*, *12*, 455–469.

Corning, W. C., Steffy, R. A., and Lahue, R., 1974, Labels and realities in psychiatric classification: A polythetic strategy, in preparation.

Corning, W. C., and Von Burg, R., 1968, Behavioral and neurophysiological investigations of *Limulus polyphemus*, *in* "Neurobiology of Invertebrates" (J. Salanki, ed.), Plenum Press, New York.

Dyal, J. A., 1971, Transfer of behavioral bias: Reality and specificity, *in* "Chemical Transfer of Learned Information" (E. J. Fjerdingstad, ed.), American Elsevier, New York.

Evans, S. M., 1969, Habituation of the withdrawal response in nereid polychaetes. I. The habituation process in *Nereis diversicolor*, *Biol. Bull.* (Woods Hole), *137*, 105–117.

Fisher, A. E., 1956, Maternal and sexual behavior induced by intracranial chemical stimulation, *Science* (Wash. D.C.), *124*, 228–229.

Fjerdingstad, E. J. (ed.), 1971, "Chemical Transfer of Learned Information," American Elsevier, New York.

Flavell, J. H., 1963, "The Developmental Psychology of Jean Piaget," Van Nostrand, New York.

Franzisket, L., 1963, Characteristics of instinctive behavior and learning in reflex activity of the frog, *Anim. Behav.*, *11*, 318–334.

Gaito, J., 1972, "Macromolecules and Behavior" (2nd ed.), Appleton-Century-Crofts, New York.

Gardner, L. E., 1968, Retention and overhabituation of a dual-component response in *Lumbricus terrestris*, *J. Comp. Physiol. Psychol.*, *66*, 315–318.

Groves, P. M., and Thompson, R. F., 1970. Habituation: A dual-process theory, *Psychol. Rev.*, *77*, 419–450.

Haigler, H. J., and Von Baumgarten, R. J., 1972, Facilitation of excitatory post-synaptic potentials in the giant cell in the left pleural ganglion of *Aplysia californica*, *Comp. Biochem. Physiol.*, *41A*, 7–16.

Harden, C. M., 1973, Behavioral modification of *Stentor coeruleus*, Reprinted in Vol. 1, Chapter 2, pp. 67–72.

Hazlett, B. A., 1971, Influence of rearing conditions on initial shell entering behavior of a hermit crab (*Decapoda paguridea*), *Crustaceana* (Leiden), *20*, 167–170.

Held, R. M., 1965, Plasticity in sensory-motor systems. *Sci. Am.*, *220*, 84–94.

Hess, W. R., 1957, "The Functional Organization of the Diencephalon," Grune & Stratton, New York.

Hirsch, H. V. B., 1972, Role of function in the development and maintenance of the cat visual system, *Neurosci. Res. Program Bull.*, *10*, 291–293.

Holmgren, B., and Frenk, S., 1961, Inhibitory phenomena and habituation at the neuronal level, *Nature* (Lond.), *192*, 1294–1295.

Horn, G., and Rowell, C. H. F., 1968, Medium and long-term changes in the behavior of visual neurones in the tritocerebrum of locusts, *J. Exp. Biol.*, *49*, 143–169.

Horridge, G. A., 1959, Analysis of the rapid responses of *Nereis* and *Harmothoe* (Annelida), *Proc. R. Soc. Lond. B. Biol. Sci.*, *150*, 245–262.

Horridge, G. A., 1962, Learning leg position by the ventral nerve cord in headless insects, *Proc. R. Soc. Lond. B. Biol. Sci.*, *157*, 33–52.

Horridge, G. A., 1968, "Interneurons," Freeman, San Francisco.

Hoyle, G., 1970, Cellular mechanisms underlying behavior-neuroethology, *Adv. Insect Physiol.*, *12*, 349–444.

Hubel, D. H., and Wiesel, T. N., 1963, Receptive fields of cells in striate cortex of very young, visually inexperienced kittens, *J. Neurophysiol.*, *26*, 994–1002.

Hyden, H., 1967, Biochemical changes accompanying learning, *in* "The Neurosciences" (G. C. Quarton, T. Melnechuk, and F. O. Schmitt, eds.), Rockefeller University Press, New York.

Jensen, D. D., 1967, Polythetic operationism and the phylogeny of learning, *in* "Chemistry of Learning" (W. C. Corning and S. C. Ratner, eds.), Plenum Press, New York.

John, E. R., 1967, "Mechanisms of Memory," Academic Press, New York

John, E. R., 1972, Switchboard versus statistical theories of learning and memory, *Science* (Wash., D.C.), *177*, 850–864.

John, E. R., and Killam, K. F., 1960, Electrophysiological correlates of differential approach-avoidance conditioning in the cat, *J. Nerv. Ment. Dis.*, *131*, 183.

Kandel, E. R., Castellucci, V., Pinsker, H., and Kupferman, I., 1970, The role of synaptic plasticity in the short-term modification of behavior, *in* "Short-term Changes in Neural

Activity and Behaviour" G. Horn and R. A. Hinde, eds., Cambridge University Press, Cambridge, England.

Kandel, E. R., and Tauc, L., 1965, Mechanisms of heterosynaptic facilitation in the giant cell of the abdominal ganglion of *Aplysia depilans, J. Physiol.* (Lond.), *181*, 28–47.

Kennedy, D., Selverston, A. I., and Remler, M. P., 1969, Analysis of restricted neural networks, *Science* Wash., D.C., *164*, 1488–1495.

Kerkut, G. A., Oliver, G. W. C., Rick, J. T., and Walker, R. J., 1970, The effects of drugs on learning in a simple preparation, *Comp. Gen. Pharmacol.*, *1*, 437–483.

Kinastowski, W., 1963*a*, Der Einfluss der mechanischen Reize auf die Kontraktilitat von *Spirostomum ambiguum, Acta Protozool.*, *1*, 201–222.

Kinastowski, W., 1963*b*, Das Problem "des lernes" bei *Spriostomum ambiguum, Acta Protozool.*, *1*, 233–236.

Krasne, F. B., 1969, Excitation and habituation of the crayfish escape reflex: The depolarizing response in lateral giant fibres of the isolated abdomen, *J. Exp. Biol.*, *50*, 29–46.

Krasne, F. B., 1973, Learning in Crustacea, in "Invertebrate Learning," Vol. 2 (W. C. Corning, J. A. Dyal, and A. O. D. Willows, eds.), Plenum Press, New York.

Krasne, F. B., and Roberts, A., 1967, Habituation of crayfish escape response during release from inhibition induced by picrotoxin, *Nature* (Lond.), *215*, 769–770.

Krasne, F. B., and Woodsmall, K. S., 1969, Waning of the crayfish escape response as a result of a repeated stimulation, *Anim. Behav.*, *17*, 416–424.

Kristan, W. B., Jr., 1971, Plasticity of firing patterns in neurons of *Aplysia* pleural ganglion, *J. Neurophysiol.*, *34*, 321–338.

Kuenzer, P. P., 1958, Verhaltenphysiologische Untersuchungen über das Zucken des Regenwurms, *Z. Tierpsychol.*, *15*, 31–49.

Lahue, R., 1974, Habituation characteristics and mechanisms in the abdominal ganglia of *Limulus polyphemus*, unpublished doctoral dissertation, University of Waterloo, Waterloo, Ontario, Canada.

Lahue, R., and Corning, W. C., 1971*a*, Habituation in *Limulus* abdominal ganglia, *Biol. Bull.* Woods Hole, *140*, 427–439.

Lahue, R., and Corning, W. C., 1971*b*, Plasticity in *Limulus* abdominal ganglia: An exercise in paleopsychology, *Can. Psychol.*, *12*, Suppl. 2, 193–194.

Lahue, R., and Corning, W. C., 1973*a*, Incremental and decremental processes in *Limulus* ganglia: Stimulus frequency and ganglion organization influences, *Behav. Biol.*, *8*, 637–653.

Lahue, R., and Corning, W. C., 1973*b*, unpublished observations.

Lashley, K. S., 1950, In search of the engram, *Symp. Soc. Exp. Biol.*, *4*, 454–482.

Lettvin, J. Y., Maturana, H. R., McCulloch, W. S., and Pitts, W. H., 1959, What the frog's eye tells the frog's brain, *Proc. Inst. Radio Engrs.*, *47*, 1940–1951.

Levi-Montalcini, R., and Chen, J. S., 1969, *In vitro* studies of the insect embryonic nervous system, *in* "Cellular Dynamics of the Neuron" (S. H. Barondes, ed.), Academic Press, New York.

Luco, J. V., 1971, A study of memory in insects, *in* "Research in Physiology" (F. F. Kao, K. Koizumi, and M. Vasalle, eds.), Aulo Gaggi, Bologna.

Morrell, R., 1961, Effect of anodal polarization on the firing pattern of single cortical cells, *Ann. N.Y. Acad. Sci.*, *92*, 860–876.

Moruzzi, G., and Magoun, H. W., 1949, Brain stem reticular formation and activation of the EEG, *Electroencephalogr. Clin. Neurophysiol.*, *1*, 455–473.

Olds, J., 1963, Mechanisms of instrumental conditioning, *in* "The Physiological Basis of Mental Activity" (R. Hernandez-Peon, ed.), American Elsevier, New York.

Palka, J., 1969, Discrimination between movements of eye and object by visual interneurones of crickets, *J. Exp. Biol.*, *50*, 723–732.

Peters, H., 1932, Experimente über die Orientierung der Kreuzspinne *Epeira diademata* Cl. im Netz, *Zool. Jahrb. Abt.*, *51*, 239–288.

Pilowsky, I., Levine, S., and Boulton, D. M., 1969, The classification of depression by numerical taxonomy, *Brit. J. Psychiatr.*, *115*, 937–945.

Pinsker, H., Kupferman, I., Castellucci, V., and Kandel, E., 1969, Habituation and dishabituation of the gill-withdrawal reflex in *Aplysia*, *Science* (Wash., D.C.), *167*, 1740–1742.

Pumphrey, R. J., and Rawdon-Smith, A. F., 1939. Synaptic transmission of nerve impulses through the last abdominal ganglion of the cockroach. *Proc. Roy. Soc. Lond. Biol. Sci.*, *122*, 106–118.

Ranney Mize, R., and Murphy, E. H., 1973, Selective visual experience fails to modify receptive field properties of rabbit striate-cortex neurons, *Science* (Wash., D.C.), *180*, 320–323.

Rasch, E., Swift, H., Riesen, A. H., and Chow, K. L., 1961, Altered structure and composition of retinal cells in dark reared mammals, *Exp. Cell Res.*, *25*, 348–363.

Ratner, S. C., 1972, Habituation and retention of habituation in the leech (*Macrobdella decora*), *J. Comp. Physiol. Psychol.*, *81*, 115–121.

Razran, E., 1971, "Mind in Evolution—An East-West Synthesis of Learned Behavior and Cognition," Houghton Mifflin, Boston.

Roberts, M. B. V., 1962, The giant fibre reflex of the earthworm, *Lumbricus terrestris* L. I. The rapid response, *J. Exp. Biol.*, *39*, 219–227.

Rosenzweig, M. R., Bennett, E. L., and Diamond, M. C., 1972, *in* "Macromolecules and Behavior" (2nd Ed.) (J. Gaito, ed.), Appleton-Century-Crofts, New York.

Rowell, C. H. F., 1971, Variable responsiveness of a visual interneurone in the free-moving locust, and its relation to behavior and arousal, *J. Exp. Biol.*, *55*, 727–747.

Rushforth, N. B., 1965, Behavioral studies of the coelenterate *Hydra pirardi* Brien, *Anim. Behav.*, *13*, Suppl. 1, 30–42.

Rushforth, N. B., 1967, Chemical and physical factors affecting behavior in *Hydra*, *in* "Chemistry of Learning" (W. C. Corning and S. C. Ratner, eds.), Plenum Press, New York.

Rushforth, N. B., 1973, Behavioral modification in coelenterate learning, *in* "Invertebrate Learning," Vol. 1 (W. C. Corning, J. A. Dyal, and A. O. D. Willows, eds.), Plenum Press, New York.

Rushforth, N. B., Burnett, A. L., and Maynard, R., 1963, Behavior in *Hydra*. Contraction responses of *Hydra pirardi* to mechanical and light stimuli; *Science* (Wash., D.C.), *139*, 760–761.

Schapiro, S., and Vukovich, K. R., 1970, Early experience effect upon cortical dendrites: A proposed model for development, *Science* (Wash., D.C.), *167*; 292–294.

Sluckin, W., 1965, "Imprinting and Early Learning," Aldine, Chicago.

Sokal, R. R., 1966, Numerical taxonomy, *Sci. Am.*, *215*, 106.

Strauss, J. S., 1974, Classification by cluster analysis, *in* "International Pilot Study of Schizophrenia," Vol. 1, in Press.

Strauss, J. S., Bartko, J. J., and Carpenter, W. T., 1973, The use of clustering techniques for the classification of psychiatric patients, *Brit. J. Psychiatr.*, *122*, 531–540.

Szlep, R., 1952, On the plasticity of instinct of a garden spider (*Aranea diadema* L.) construction of a cobweb, *Acta Biol. Exp.* (Warsaw), *16*, 5–24.

Thompson, R. F., and Spencer, W. A., 1966, Habituation: A model phenomenon for the study of neuronal substrates of behavior, *Psychol. Rev.*, *73*, 16–43.

Thorpe, W. H., 1939*a*, Further studies on olfactory conditioning in a parasitic insect. The nature of the conditioning process, *Proc. R. Soc. Lond. Biol. Sci., 126,* 379–397.

Thorpe, W. H., 1939*b*, Further studies of pre-imaginal olfactory conditioning in insects, *Proc. R. Soc. Lond. B. Biol. Sci., 127,* 424–433.

Von Burg, R., and Corning, W. C., 1970, Cardioregulatory properties of the abdominal ganglion in *Limulus. Can. J. Physiol. Pharmacol. 45,* 333–346.

Vowles, D. M., 1961, Neural mechanisms in insect behavior, *in* "Current Problems in Animal Behavior" (W. H. Thorpe, and O. L. Zangwill, eds.), Harvard University Press, Cambridge.

Wawrzynczyk, S., 1937, Badania nad pamiecia *Spirostomum ambiguum major, Acta Biol. Exp.* (Warsaw), *11,* 57–77.

Wickelgren, B. C., 1967*a*, Habituation of spinal motoneurons, *J. Neurophysiol., 30,* 1404–1423.

Wickelgren, B. C., 1967*b*, Habituation of spinal interneurons, *J. Neurophysiol., 30,* 1424–1438.

Wiersma, C. A. G., and Yanagisawa, K., 1971, On types of interneurons responding to visual stimulation present in the optic nerve of the rock lobster, *Panulirus interruptus, J. Neurobiol., 2,* 291–309.

Willows, A. O. D., 1969, Neuronal network triggering a fixed action pattern, *Science* (Wash., D.C.), *166,* 1549–1551.

Wine, J. J., 1973, Invertebrate central neurons: Orthograde degeneration and retrograde changes after axotomy, *Exp. Neurol., 38,* 157–169.

Wood, D. C., 1970*a*, Parametric studies of the response decrement produced by mechanical stimuli in the protozoan, *Stentor coeruleus, J. Neurobiol., 1,* 345–360.

Wood, D. C., 1970*b*, Electrophysiological studies of the protozoan, *Stentor coeruleus, J. Neurobiol., 1,* 363–377.

Wood, D. C., 1971, Electrophysiological correlates of the response decrement produced by mechanical stimuli in the protozoan, *Stentor coeruleus, J. Neurobiol., 2,* 1–11.

Wood, D. C., 1972, Generalization of habituation between different receptor surfaces of *Stentor, Physiol. Behav., 9,* 161–165.

Zucker, R. S., Kennedy, D., and Selverston, A. I., 1971, Neuronal circuit mediating escape responses in crayfish, *Science* (Wash., D.C.), *173* 645–650.

Appendix A

LEARNING IN BACTERIA, FUNGI, AND PLANTS

PHILIP B. APPLEWHITE[1]

Division of Natural Sciences
State University of New York
College at Purchase, New York

There has always been some interest in demonstrating learning in bacteria, fungi, or plants if for no other reason than to extend the phenomenon beyond animals to other kingdoms. From strictly a comparative psychology viewpoint, this would be most interesting. However, an extension of learning to bacteria such as *Escherichia coli* or fungi such as *Phycomyces* could be most important, as these organisms lend themselves to powerful genetic analysis of any behavior they possess. Thus far, genetic analysis at the molecular level has not really been feasible with the animals commonly used in learning studies with the current exception of *Drosophila,* which shows conditioned behavior (Quinn *et al.*, 1974). Presumably, what one would like to find is a mutant strain that is either deficient or proficient in some learning experiment relative to a wild-type strain. If it could be shown for example, that only one gene was involved, unrelated to a nutritional requirement necessary for normal cell maintenance, then considerable progress could be made in elucidating the memory mechanisms. While the specific gene involved could be under the control of some other regulatory gene that was at the center of memory coding, the gene would still be an important part of the memory system. Such analysis is being used now in the study of endogenous circadian clocks with mutants of *Chlamydomonas* (Bruce, 1972) and *Neurospora* (Feldman and Hoyle, 1973) in attempts to locate the oscillator.

[1] Visiting Fellow in Biology, Yale University, New Haven, Connecticut

The behavior of bacteria has been studied since the last century and its early history has been reviewed by Jennings (1906). None of the experiments mentioned, involving mostly taxis studies, dealt with obtaining learning in bacteria. Recent conference proceedings on the photobiology (Halldal, 1970) and behavior (Pérez-Miravete, 1973) of microorganisms also have not brought forward any experiments on bacterial learning, although they have presented useful information on behavior that could serve as a background for the designing of learning experiments. Bacteria are clearly responsive, with motility changes to a wide range of stimuli such as light, temperature, chemicals, and electricity, but probably not gravity (Brown, 1971). There seems to be no *a priori* reason why stimuli could not be chosen that would fit into a classical conditioning paradigm and used on bacteria. Because of their small size, the experimental apparatus would have to be scaled down in size and used with microscopes, but should present no great difficulty. While the *E. coli* bacterium is 2–3 μm in length (0.5 μm wide), some bacteria are relatively large. Members of the genus *Spirillum* (found in stagnant water) can reach 40 μm in length; *Thiospirillum* (stagnant water containing hydrogen sulfide) can reach 100 μm in length; and *Spirochaeta plicatilis* (fresh water) can reach lengths of 500 μm (Breed *et al.* 1957). For comparison, the smallest multicelled organisms are only about 50 μm in length.

Some recent protozoan learning-experiment designs might serve as models in the designing of bacterial learning experiments. The design used by Bergstrom (1968) on *Tetrahymena* involved pairing light flashes with electric shocks, then exposing the protozoa to light and measuring whether they then avoided the light. This was done by their being kept in the dark between light–shock pairings. After the pairings, a screen with circular holes was placed under their transparent chamber with light directed underneath. It was reasoned that if the light–shock pairings were effective, the *Tetrahymena* would avoid the light coming through the holes and swim away to the area between the holes. While this behavior could not be replicated (Applewhite *et al.*,1971), it would be something to try with bacteria. Other protozoan learning experiments amenable to bacteria involve tube escape behavior (Hanzel and Rucker, 1972; Bennett and Francis, 1972). In these experiments, *Paramecia* or *Stentor* were sucked up a small capillary tube, the tip of which was kept in a pool of media, and the times when they swam back out into the pool were recorded. The procedure was repeated several times, and eventually they took less time to swim out. The *Paramecia* results were confirmed but explanations other than learning were suggested (Applewhite and Gardner, 1973), as they were also by the authors of the *Stentor* experiments. The larger bacteria could be used here and any changes in escape time over trials could be measured. These bacteria could also be tested for habituation responses. With a micromanipulator, a probe could be touched to one end of a bacterium to determine if it swam away from the tip with an avoidance response. If it eventually ceased this behavior and swam by the probe with no avoidance response, we would have an interesting start toward studying adaptive behavior in procaryotes.

While scaling down vertebrate-type learning experiments for use with single-celled organisms may seem quite reasonable, it just may be that this is why conditioning has not been clearly demonstrated in this group. It may be necessary in eliciting learning from them to use stimuli appropriate to their own natural environment. It is not certain that electrical shocks or flashes of light are ever registered for most unicellular organisms unless they noticeably show a definite response, such as contraction, to stimuli as shown by the protozoa *Stentor, Spirostomum, Vorticella, Zoothamnium,* and *Carchesium,* for example. Consequently, experiments utilizing chemotaxis in bacteria, which look at behavior without imposing stimuli of a kind foreign to the organism's environment, seem most promising in attempts to find learning and memory. One way in which chemotaxis is measured is by the placement of the desired attractant liquid in a capillary tube sealed at one end, with the open end placed in a suspension of bacteria. After a wait of up to 1 hr, the tube is withdrawn, the liquid in it containing the attracted bacteria is plated, and the colonies are counted the next day (Adler, 1973). The literature in this area is growing rapidly and a review of it goes beyond the subject of this paper. The work has been done with *E. coli* and *Salmonella,* with both wild and mutant-nonchemotactic strains. They have been studied in their response to a variety of attractants, including oxygen (Adler, 1966), sugars (Hazelbauer and Adler, 1971), and amino acids (Mesibov and Adler, 1972). Because of the large amount of information available on the biochemical genetics of these bacteria, it is hoped that these studies of chemotaxis will lead to a better understanding of the molecular basis of sensory reception in the Metazoa. As Koshland (1974) has stated, "the analogy to pheromones in insects and to taste and odor in mammals is impressive."

To date, no conditioning experiments have been done with the chemotactic responses, but the question has been posed, "Can bacteria 'learn' to swim toward or away from a chemical?" (Adler, 1966). However, an experiment has been done with *Salmonella* that indicates that a memory of the bacterial environment exists (Macnab and Koshland, 1972). When the bacteria in a solution containing 1 mM serine (an attractant) were suddenly mixed in with culture media containing no serine (resulting in a quick fall of serine concentration in the mixture from 1 mM to 0.24 mM), they showed erratic swimming behavior typical of their reaction to meeting a gradient of lesser concentration. However, the mixing took place so fast (less than 1 sec) that no gradient in concentration was in fact present. The fact the bacteria showed a swimming response as though they were going down a gradient indicated that they possessed some temporal memory that compared past and present concentrations. After 12 sec, their swimming behavior became normal—the memory had decayed. If the bacteria were in their culture media and then were suddenly mixed with a 1-mM serine concentration (resulting in a quick rise in the serine concentration in the mixture from zero to 0.76 mM), they showed supercoordinated swimming behavior typical of their reaction to meeting a gradient of higher concentration. Again, there was no concentration gradient present—only a change in concentration. Their swimming response in the mixture was as though they were going up a gradient—demonstrating a memory of their past concentration. This memory decayed to normal swimming behavior within 5 min. Control experiments mixing the bacteria with the same media they were in produced no changes in normal swimming behavior. This kind of memory has also been demonstrated in a similar experiment utilizing repellent substances rather than attractants (Tsang *et al.*, 1973). The chemoreceptors that react to the chemical stimuli may change the electrical properties of the bacterial cell membrane, thereby affecting the flagella (which are in contact with it) and therefore the swimming behavior (Hazelbauer and Adler, 1971). As the authors point out, it remains to be determined whether this form of memory has any relation to memory in higher organisms, but it certainly is time-dependent information, stored and then used by the organism. It has been characterized as useful memory (Macnab and Koshland, 1974). Nevertheless, it marks the first promising evidence of memory in bacteria. What one would hope for is that explanations of the memory here would suggest what specifically to look for in the biochemical basis of memory in higher organisms.

The fungi have turned out to be quite useful in the study of behavior not only, as mentioned, in biological clock research but also in sensory physiology (Bergman *et al.*,1969). The sporangiophore of *Phycomyces*, in particular, has a growth rate that can be controlled by gravity, stretch, light, and the presence of solid objects very close by. The latter two stimuli have been used to study habituation of the growth response (Ortega and Gamow, 1970). The sporangiophore had a normal rate of growth under a constant-intensity light source, but when the intensity was increased and maintained, the rate of growth increased within minutes. Within a few more minutes, the growth rate decreased to the normal growth rate. The sporangiophore had to adapt to this increased light intensity before responding to another increase in light intensity. It was habituated to this light intensity because upon subsequent presentation of it no increase in growth rate resulted—an increased light intensity was now necessary. To eliminate the explanation that depletion of growth substances caused the habituation, another experiment was performed. After habituation to a light intensity of a certain intensity, two glass cover slips were placed around the sporangiophore but not touching it. This produced an avoidance response (it is not known how it senses the nearness of objects such as the cover slip, possibly from changes in vapor pressure around the fungi), causing an increase in vertical growth rate. Therefore fatigue (the depletion of chemical substances) does not appear to be an explanation of habituation to the light growth response. Whether habituation is a form of learning or not has been discussed in these volumes, but it is something that cannot be unequivocally resolved now (Applewhite, 1973). In any case studying habituation in this organism seems as valid as studying it in other organisms in attempts to understand better the chemistry of memory. The early literature mentioned the behavior of fungi with regard to their response to stimuli, but no references to learning of any kind were made (Warden *et al.,* 1940).

Warden *et al.* (1940) did mention some of the early observations on adaptation (habituation) to stimulation in the higher plants, however. Pfeffer (1873) wrote about the adaptation of *Mimosa* leaflets to repeated stimulation. Under constant repeated stimulation the leaflets first closed and then gradually opened fully and did not close under the stimulation. He also mentioned the case of the carnivorous plant *Drosera* (sundew), whose tentacle movement became habituated to repeated touches (Pfeffer, 1906). Darwin (1875) noted that when a thin stick was rubbed on tendrils of *Passiflora gracilis* (passion flower) they coiled within several seconds, then would uncoil sometime later. Rubbing them everytime the tendrils uncoiled produced a response 21 times in a row over a period of 54 hr. After this, further stimuli failed to produce a response. Bose (1906, 1913, 1928) studied habituation of the petiole-falling response in *Mimosa.* When the stem (petiole) was touched or electrically shocked, it fell, rising again many minutes later. Under constant repeated electrical or mechanical stimulation, the petiole first fell, then eventually rose. Only after a suitable rest period of no stimulation did subsequent stimulation cause a fall of the petiole. Bose also looked at the effect of a variety of chemicals (including anesthetics) upon the leaflets sensitivity to stimulation, as did Wallace (1931 *a,b,c*) some years later. Some of the results were confirmed and extended and showed that drugs like curare and xylocaine prevented leaflets from closing to stimulation (Applewhite, 1972*b*).

More recently, Holmes *et al.* (1965, 1966) have studied habituation of leaflet closure in *Mimosa*, using either drops of water or brush strokes as leaflet stimuli. Repeated presentations of stimuli eventually caused the leaflets to remain open during subsequent stimulation. A more detailed study of habituation in *Mimosa* looked at the effects of different intensities of stimulation, different intertrial intervals, and a comparison of electrical with mechanical stimulation (Applewhite, 1972*a*). Several leaflets were excised from a plant, floated in water, and studied *in vitro*. With a constant-intensity mechanical stimulus presented to the leaflets every 2 sec, the leaflets closed, but after about 13 min of this stimulation they became fully open and were insensitive to stimulation. Increasing the intertrial interval to 15 sec produced opening after 29 min. When the intensity of stimulation was increased, with the intertrial interval held constant, the leaflets opened after a much longer period of stimulation. As with many animal experiments, in *Mimosa* the longer the intertrial interval and the more intense the stimulus, the longer it takes to produce habituation. Dishabituation was not demonstrated in this system. An electrical shock presented through the water produced leaflet closure, and it was shown that constant, repeated shocks, like mechanical stimulation, produced habituation of leaflet closure. Furthermore, leaflets habituated to electrical shocks still closed and habituated immediately thereafter to a series of mechanical stimuli, and vice versa. The first mode of stimulation also had no effect on the time it took for habituation to occur with the second mode of stimulation. For example, it took an average of about 9 min for a series of electrical shocks to produce leaflet habituation, whether or not the leaflets had been previously habituated with mechanical stimulation.

Research on the conditioning of plants has been limited to *Mimosa*.[1] Holmes *et al*. (1965, 1966) paired a light touch conditioned stimulus (CS), which produced little or no leaflet closure, with an electrical shock unconditioned stimulus (UCS), which followed soon thereafter, and produced leaflet closure. No conditioning was produced as the light touch alone never produced widespread leaflet closure. They also failed to produce conditioning when leaflets were lightly brushed with a brush (CS) followed by a strong jarring of the leaflets (UCS), as the CS never produced a response similar to that of the UCS. Haney (1969) tried conditioning leaflets by using as the CS either light or darkness, paired with touch (UCS) that produced leaflet closure. Giving the plants darkness alone did not produce the leaflet closure after training, but in some cases after training, the stimulus of light alone did produce leaflet closure. However, adequate controls, such as random light and touch stimuli, light only or touch only, were not run. Furthermore, the findings could not be replicated (Levy *et al*., 1970). A similar experiment by Armus (1970) was claimed to have produced conditioning using darkness as the CS and touch as the UCS. But again, necessary controls were not run as in the Haney experiment and interpretation is not clear. It has been shown, for example, that if open leaflets are placed in the dark for at least 5 min, then given light, the light alone will produce rapid leaflet closure independent of any other previously paired stimulus (Applewhite and Gardner, 1971). Thus the choice of con-

[1] I have heard indirectly for many years that attempts have been made to condition Venus's-flytrap, but thus far nothing has appeared in the literature I have seen.

trols is important to the decision of whether conditioning has been achieved. In view of the many kinds of plants showing rapid movements (Sibaoka, 1969), there is certainly a wide range of conditioning and habituation experiments that could be designed, and should be.

From a behavioral standpoint, it must be concluded that conditioning has not been clearly demonstrated in single-celled organisms, in Metazoa, which lack a nervous system, in fungi, or in plants. Since so little research has been done with these organisms, it would be premature to suggest that a nervous system is necessary for "learning" to occur. It becomes presumptuous to say the organisms above do not "need" to learn, either because their generation times are so short or because they could not quickly move to another location to take advantage of the learning, but for now we have no stronger conclusion.

REFERENCES

Adler, J., 1966, Chemotaxis in bacteria, *Science* (Wash., D.C.), *153*, 708.

Adler, J., 1973, A method for measuring chemotaxis and use of the method to determine optimum conditions for chemotaxis by *E. coli*, *J. Gen. Microbiol.*, *74*, 77.

Applewhite, P. B., 1972*a*, Behavioral plasticity in the sensitive plant, *Mimosa*, *Behav. Biol.*, *7*, 47.

Applewhite, P. B., 1972*b*, Drugs affecting sensitivity to stimuli in the plant *Mimosa* and the protozoan *Spirostomum*, *Physiol. Behav.*, *9*, 869.

Applewhite, P. B., 1973, Habituation in the protozoan *Spirostomum* and problems of learning, *in* "Behavior of Micro-organisms" (A. Pérez-Miravete, ed.), Plenum Press, New York.

Applewhite, P. B., and Gardner, F., 1971, Rapid leaf closure of *Mimosa* in response to light, *Nature* (Lond.), *233*, 279.

Applewhite, P. B., and Gardner, F., 1973, Tube-escape behavior of *Paramecia*, *Behav. Biol.*, *9*, 245.

Applewhite, P. B., Gardner, F., Foley, D., and Clendenin, M., 1971, Failure to condition *Tetrahymena*, *Scand. J. Psychol.*, *12*, 65.

Armus, H. L., 1970, Conditioning of the sensitive plant *Mimosa pudica*, *in* "Comparative Psychology: Research in Animal Behavior" (M. R. Denny and S. C. Ratner, eds.), Dorsey Press, Homewood, Ill.

Bennett, D. A., and Francis, D., 1972, Learning in *Stentor*, *J. Protozool.*, *19*, 484.

Bergman, K., Burke, P. V., Cerda-Olmedo, E., David, C. N., Delbruck, M., Foster, K. W., Goodell, E. W., Heisenberg, M., Meissner, G., Zalokar, M., Dennison, D. S. and Shropshine, W. Jr., 1969, Phycomyces, *Bacteriol. Rev.*, *33*, 99.

Bergstrom, S. R., 1968, Induced avoidance behaviour in the protozoa *Tetrahymena*, *Scand. J. Psychol.*, *9*, 215.

Bose, J. C., 1906, "Plant Response," Longmans, London.

Bose, J. C., 1913, "Researches on Irritability of Plants," Longmans, London.

Bose, J. C., 1928. "The Motor Mechanism of Plants," Longmans, London.

Breed, R. S., Murray, E. G. D., and Smith, N. R., 1957, "Bergey's Manual of Determinative Bacteriology," Williams and Wilkins, Baltimore.

Brown, A. H., 1971, The organism and gravity: An introduction, *in* "Gravity and the Organism" (S. A. Gordon and M. J. Cohen, eds.), University of Chicago Press, Chicago.

Bruce, V. G., 1972, Mutants of the biological clock in *Chlamydomonas reinhardii, Genetics, 70,* 537.

Darwin, C., 1875, "The Movements and Habits of Climbing Plants," J. Murray, London.

Feldman, J. F. and Hoyle, M., 1973, Isolation of circadian clock mutants of *Neurospora crassa, Biophysical Society Abstracts*, 148a.

Halldal, P., ed., 1970, "Photobiology of Microorganisms," Wiley New York.

Haney, R. E., 1969, Classical conditioning of a plant: *Mimosa pudica, J. Biol. Psychol., 11,* 5.

Hanzel, T. E., and Rucker, W. B., 1972, Trial and error learning in *Paramecia:* A replication, *Behav. Biol., 7,* 873.

Hazelbauer, G. L., and Adler, J., 1971, Role of the galactose binding protein in chemotaxis of *E. coli* toward galactose, *Nat. New Biol., 230,* 101.

Holmes, E., and Greenberg, G., 1965, Learning in plants? *Worm Runner's Digest, 7,* 9.

Holmes, E., and Yost, M., 1966, Behavioral studies in the sensitive plant, *Worm Runner's Digest, 8,* 38.

Jennings, H. S., 1906, "Behavior of the Lower Organisms," Indiana University Press, Bloomington.

Koshland, D. E., Jr., 1974, The chemotactic response as a potential model for neural systems, *in* "The Neurosciences, 3rd Study Program" (F. O. Schmitt and F. G. Worden, eds.), MIT Press, Cambridge.

Levy, E., Allen, A., Caton, W., and Holmes, E., 1970, An attempt to condition the sensitive plant *Mimosa pudica, J. Biol. Psychol., 12,* 86.

Macnab, R., and Koshland, D. E., Jr., 1972, The gradient sensing mechanism in bacterial chemotaxis, *Proc. Nat. Acad. Sci.* U.S.A., *69,* 2509.

Macnab, R., and Koshland, D. E., Jr., 1974, Persistence as a concept in the motility of chemotactic bacteria, *J. Mechanochem. Cell Motility*, in press.

Mesibov, R., and Adler, J., 1972, Chemotaxis toward amino acids in *E. coli, J. Bacteriol., 112,* 315.

Ortega, J. K. E., and Gamov, R. I., 1970, Phycomyces: habituation of the light growth response, *Science* (Wash., D.C.), *168,* 1374.

Pérez-Miravete, A., 1973, "Behaviour of Micro-organisms," Plenum Press, New York.

Pfeffer, W., 1873, "Physiologische Untersuchungen," Leipzig.

Pfeffer, W., 1906, "The Physiology of Plants," Vol. 3, Clarendon Press, Oxford.

Quinn, W. G., Harris, W. A., and Benzer, S., 1974, Conditioned behavior in *Drosophila melanogaster, Proc. Nat. Acad. Sci.* U.S.A., *71,* 708.

Sibaoka, T., 1969, Physiology of rapid movements in higher plants, *Annu. Rev. Plant Physiol., 20,* 165.

Tsang, N., Macnab, R., and Koshland, D. E., Jr., 1973, Common mechanism for repellants and attractants in bacterial chemotaxis, *Science* Wash., D.C., *181,* 60.

Wallace, R. H., 1931*a*. Studies on the sensitivity of *Mimosa pudica*. I. The effect of certain animal anesthetics upon sleep movements, *Am. J. Bot., 18,* 102.

Wallace, R. H., 1931*b*. Studies on the sensitivity of *Mimosa pudica*. II. The effect of animal anesthetics and certain other compounds upon seismonic activity, *Am. J. Bot., 18,* 215.

Wallace, R. H., 1931*c*, Studies on the sensitivity of *Mimosa pudica*. III. The effect of temperature, humidity and certain other factors upon seismonic sensitivity, *Am. J. Bot., 18,* 288.

Warden, C. J., Jenkins, T. N., and Warner, L. H., 1940, "Comparative Psychology," Vol. 2, Ronald, New York.

APPENDIX B

PROGRESSIONS: BIBLIOGRAPHY OF RECENT PUBLICATIONS[1]

SONJA ZIGANOW AND W. C. CORNING

Department of Psychology
University of Waterloo
Waterloo, Ontario, Canada

Protozoa (Vol. 1)

Applewhite, P. B., 1972, Drugs affecting sensitivity to stimuli in the plant *Mimosa* and the protozoan *Spirostomum*, *Physiol. Behav.*, *9*, 869–871.

There is a close similarity in the effects of drugs on the sensitivity to mechanical stimulation in the sensitive plant *Mimosa* and the protozoan *Spirostomum*. Certain chemicals also affect the rate of opening to light and closure to dark in *Mimosa*. The chemicals would appear to affect membrane permeability within the two organisms by similar processes.

Applewhite, P. B., and Gardner, F. T., 1973, Tube-escape behavior of paramecia, *Behav. Biol.*, *9*, 245–250.

Paramecia sucked up into a small capillary tube eventually took less and less time to escape from it. This change in escape time over trials could be eliminated if the tubes were cleaned in a prescribed manner before use and the organisms were adapted to the process of being sucked up before the experiments began.

Hanzel, T. E., and Rucker, W., 1971, Escape training in paramecia, *J. Biol. Psychol.*, *73*, 24–28.

For ten trials, single paramecia were sucked into a capillary tube and allowed to escape back into a drop of their own culture medium in an attempt to demonstrate escape learning in single-celled organisms. Neither variations in intertrial interval (massed

[1] The authors wish to thank those contributors to various chapters who took the time to provide us with additional references.

practice or 1 min) nor variations in the inner diameter of the tube (0.45 mm or 0.63 mm) affected escape speed, which increased reliably over the ten training trials. A measure of general activity was taken just before and just after escape training. Although activity increased (contrary to French, 1940), there was no significant interaction between the increase in activity and the rate of increase in escape speed. Control animals that spent 3 min in the tube prior to training, easily developed rapid escape. These results and previous work lend support to the hypothesis that single-celled animals are capable of demonstrating changes in behavior that in higher animals would be described as learning.

Hanzel, T. E., and Rucker, W. B., 1972, Trial and error learning in paramecia: A replication, *Behav. Biol.*, *7*, 873–880.

Single paramecia were sucked into a capillary tube and allowed to escape back into a drop of their culture. Escape speed increased in early trials and then stabilized, thus replicating French's 1940 results and supporting his hypothesis of learning in paramecia. Further, neither intertrial interval nor habituation to the drop changed the rate at which escape improved, though habituation increased escape speed over all trials.

Huber, J. C., 1972, Speculations concerning the physiology of learning in paramecia, *J. Biol. Psychol.*, *14*, 22–26.

This article reviews evidence that attempts to show a relationship between neural-like functioning in paramecium and multineuronal systems functioning in high organisms. Two possible candidates are presented for a nervous system analogue, i.e., the neurofibril system and the cell membrane. Also presented is evidence for sensory functions of cilia. Finally, the chemical similarities in paramecium and the multineuronal system's functioning are discussed.

Huber, J. C., Rucker, W. B., and McDiarmid, C. G., 1974, Retention of escape training and activity changes in single paramecia, *J. Comp. Physiol. Psychol.*, *86*, 258–266.

Paramecia sucked individually into a capillary tube and allowed to swim back into a drop of their own culture medium escape more rapidly over successive trials (French, 1940). Hanzel and Rucker (1971, 1972) reported that neither variations in the intertrial interval (ITI), nor inner diameter of the capillary tube, nor time in the drop prior to training affected the rate at which escape speed increased over trials, and that escape speed was asymptotic within ten trials. In the present experiment, 108 subjects were trained for ten trials, and after 0, 6, 30, or 150 min were tested for retention over ten trials. During the ITIs, activity in the drop was measured and used as a covariate adjuster for escape speed. Escape speed increased reliably in the training session and was retained at all intervals. Activity was not correlated with escape speed across retention intervals. Room temperature and culture pH at running time did not interact with the retention interval.

Mrosovsky, N., 1973, Temperature and learning in poikilotherms, *J. Theor. Biol.*, *39*, 659–663.

Applewhite has suggested that learning might involve relatively temperature-independent diffusion of particles of low molecular weight, while memory decay might involve temperature-dependent active transport of those substances back to their prelearning location. This suggestion is derived from studies on the effects of temperature on learning and retention in poikilotherms. An examination of four studies that support the proposition that rates of acquisition are essentially the same at one temperature compared to another shows that in general they do not really support this proposition.

Osborn, D., Blair, H. S., Thomas, J., and Eisenstein, E. M., 1973, The effects of vibratory and electrical stimulation on habituation in the ciliated protozoan, *Spirostomum ambiguum, Behav. Biol., 8,* 655–664.

The effects of vibratory and electrical stimulation on habituation in *Spirostoma* have been investigated. Vibratory stimuli delivered at the rate of one stimulus per 10 sec for 10 min led to a decrement in the probability of contracting. Recovery occurred within 30–45 min. Two-millisecond biphasic electric shocks, which elicited contraction, did not cause a decrement in the probability of contracting when presented at a rate of once per 10 sec for 10 min. The interaction of the two modes of stimulation on the same animal was investigated. Electrical stimulation led to a decrement in the probability of contracting to a vibratory stimulus. However, vibratory stimulation did not alter the probability of contracting to electrical stimulation.

Osborn, D., Hsung, J. C., and Eisenstein, E. M., 1973, The involvement of calcium in contractility in the ciliated protozoan, *Spirostomum ambiguum, Behav. Biol., 8,* 665–677.

Parnas, M. R. I., and Nevo, A. C., 1969, Extensibility and tensile strength of the stalk "muscle" of *Carchesium* sp., *Exp. Cell Res., 54,* 69–76.

Patterson, D. J., 1973, Habituation in a protozoan, *Vorticella convallaria, Behaviour, 45,* 304–311.

The response of *Vorticella convallaria* to electrical and mechanical stimuli was investigated. It was shown that:

1. *V. convallaria* becomes habituated independently to both mechanical and electrical stimuli if these are repeated at 10-sec intervals.
2. The habituation is more rapid to lower intensities of stimuli.
3. The form of the response curve is similar for both types of stimulus.
4. No habituation was obtained when the animals were stimulated while in deionized water. No response was obtained while they were placed in a solution of $CaCl_2$ but continuous contraction was induced in a solution of KOH. Other potassium and sodium salts in deionized water gave response curves similar to those obtained under normal conditions.

Personal communication with Dr. Patterson re error: "The top curve in Fig. I should be marked 150,000 ergs. and the bottom curve 110,000 ergs. It follows that line 3 of Results should be amended 'higher' to 'lower.'"

Rahat, M., Pri-paz, Y., and Parnas, I., 1973, Properties of stalk "muscle" contractions of *Carchesium*, sp., *J. Exp. Biol., 58,* 463–471.

Wood, D. C., 1970*c*, Electrophysiological correlates of the response decrement produced by mechanical stimuli in the protozoan, *Stentor coeruleus, J. Neurobiol., 2,* 1–11.

The response pattern of *Stentor coeruleus* has been studied by the use of repetitive mechanical stimuli that were applied to a limited body surface. The decrement in the probability of response was found to parallel that previously reported when more gross mechanical stimuli were employed. This result justified the employment of such localized stimuli in the determination of electrophysiological correlates of the response decrement. Electrophysiological records show a decrease in the amplitude of the prepotential and an increase in the latency of the response during the course of repetitive stimulation. After consideration of possible complications resultant from changes in the amplitude of the diphasic potential and the relation of the stimulus probe to the animal, it was concluded that the potential changes noted do, in fact, indicate that a change in receptor function is associated with the response decrement. This result was predicted on the basis of previously reported behavioral data.

Wood, D. C., 1972, Generalization of habituation between different receptor surfaces of *Stentor, Physiol. Behav.*, *9*, 161–165.

The protozoan *Stentor coeruleus* was repeatedly stimulated with a reproducible mechanical stimulus applied to a small portion of its cell surface. The animal's probability of contracting in response to the stimulus decreased with repeated stimulation. After this habituation had occurred, the site of stimulation was moved to another area of the cell surface, but the decrement in responsiveness had generalized to that area also. Such generalization of habituation was not resultant from the contractions observed during habituation, since contractions induced by electrical stimuli did not depress response probabilities. Contractions in *Stentor* were associated with diphasic potentials which appeared synchronously at different sites on the cell surface. These potentials appear to be propagated away from the site of mechanical stimulation at a rate sufficient to explain the speed with which the animal contracts. The directions of propagation observed require that the cell have at least two different receptor surfaces capable of independently initiating action potentials. Generalization of habituation may result from a diffusion process occurring between these receptor surfaces.

Wood, D. C., 1973*a*, Physiological correlates of habituation in *Stentor coeruleus, in* "Behavior of Micro-organisms," (A. Pérez-Miravete, ed.), Plenum Press, New York.

Wood, D. C., 1973*b*, Stimulus specific habituation in a protozoan, *Physiol. Behav.*, *11*, 349–354.

Stentor coeruleus can contract in an all-or-none fashion in response to mechanical and photic stimuli. Upon repeated presentation of either of these stimuli, a decrement in the probability of eliciting a contraction is observed. However, repeated mechanical stimulation does not affect the animals' sensitivity to photic stimuli and conversely. This stimulus specificity suggests that habituation occurs in processes that are not common to the photic and mechanical stimulus–response sequences. Action potentials are shown to be produced by both stimuli. However the receptor potentials generated by photic and mechanical stimuli differ in form and in sensitivity to polarizing currents. It is concluded that the receptor potential and transduction mechanisms are not common to both forms of stimulation, and therefore one or both of these mechanisms may be changing during the course of habituation.

Behavioral Modifications in Coelenterates (Vol. 1)

Ball, E. E., 1973, Electrical activity and behavior in the solitary hydroid *Corymorpha palma*. I. Spontaneous activity in whole animals and in isolated parts, *Biol. Bull.* (Woods Hole), *145*, 223–242.

Ball, E. E., and Case, J. F., 1973, Electrical activity and behavior in the solitary hydroid, *Corymorpha palma*. II. Conducting systems, *Biol. Bull.* (Woods Hole), *145*, 243–264.

Hand, S. R., and Gobel, S., 1972, The structural organization of the septate and gap junctions of *Hydra*, *J. Cell Biol.*, *52*, 397–408.

Josephson, R. K., and Rushforth, N. B., 1973, The time course of pacemaker inhibition in the hydroid *Tubularia*, *J. Exp. Biol.*, *59*, 305–314.

Kass-Simon, G., 1972, Longitudinal conduction of contraction burst pulses from hypostome excitation loci in *Hydra attenuata*, *J. Comp. Physiol.*, *80*, 29–49.

McFarlane, I. D., 1969, Co-ordination of pedal-disk detachment in the sea anemone, *Calliactis parasitica*, *J. Exp. Biol.*, *51*, 387–396.

McFarlane, I. D., 1970, Control of preparatory feeding behavior in the sea anemone *Tedia felina*, *J. Exp. Biol.*, *53*, 211–220.

Morin, J. G., and Cooke, I. M., 1971*a*, Behavioral physiology of the colonial hydroid, *Obelia*. I. Spontaneous movements and correlated electrical activity, *J. Exp. Biol.*, *54*, 689–706.

Morin, J. G., and Cooke, I. M., 1971*b*, Behavioral physiology of the colonial hydroid, *Obelia*. II. Stimulus-initiated electrical activity and bioluminescence, *J. Exp. Biol.*, *54*, 707–721.

Rushforth, N. B., 1973, Behavior of *Hydra*, *in* "The Biology of Hydra" (A. L. Burnett, ed.), pp. 3–41, Academic Press, New York.

Rushforth, N. B., and Hofman, F., 1972, Behavioral and electrophysiological studies of *Hydra*. III. Components of feeding behavior, *Biol. Bull.* (Woods Hole), *142*, 110–131.

Spencer, A. N., 1971, Moyoid conduction in the siphonophore *Nanomia bijuga*, *Nature* (Lond.), *223*, 490–491.

Westfall, J. A., 1973*a*, Ultrastructural evidence for a granule-containing sensory-motor-interneuron in *Hydra littoralis*, *J. Ultrastruct. Res.*, *42*, 268–282.

Westfall, J. A., 1973, Ultrastructural evidence for neuromuscular systems in coelenterates, *Am. Zool.*, *13*, 237–246.

Platyhelminthes: The Turbellarians (Vol. 1)

Becker-Carus, C. von, 1970, Die Bedeutung der Tageszeit fur Sensibilitat, Reizattigung und Entscheidungsactivitat bei Planarian, *Z. Tierpsychol.*, *27*, 761–770.

Dugesia dorotocephala were found to be more sensitive to brief electrical shock in the evening and night than in the morning and at noon. Habituation was observed in the morning and evening groups. Diurnal variation in activity was examined in a Y-maze.

Coward, S. J., and Vitale-Calpe, R. O., 1971, Behaviorally significant surface specializations of the planarian, *Dugesia dorotocephala, J. Biol. Psych., 13,* 3–10.

By means of scanning and transmission electron microscopy, we have been able to describe the morphology of mucus secretion, rhabdite ejection, and epidermal specializations presumed to be sensory receptors. The probable roles of each are discussed in light of previous behavioral information.

Jones Harden, F. R., 1971, The response of the planarian *Dendrocoelum lacteum* to an increase in light intensity, *Anim. Behav., 19,* 269–276.

Dendrocoelum is said to respond to an increase in light intensity by an increase in its rate of change of direction (rcd). An infrared viewing technique was used to observe and record, on 16-mm film, the response of dark-adapted *Dendrocoelum* to sudden increase in light intensity of 1 lux. The planarians moved faster, turned through bigger angles, and turned more frequently; the response has both ortho- and klinokinetic components. The increase in the speed of gliding was maintained after the rcd had fallen back to its original level. *Dendrocoelum* reacted to intensities as low as 10^{-5} lux, and at these intensities they moved faster and turned more frequently but the mean angle of turn decreased: there was a fall in red. The results are discussed in relation to Ullyott's earlier work and shock reactions.

Kimmel, H. D., and Garrigan, H. A., 1973, Resistance to extinction in planaria, *J. Exp. Psychol., 101,* 343–347.

Planaria were classically conditioned, using 150 or 250 conditioning trials and a 2- or 4-sec conditioned stimulus-unconditioned stimulus (CS–USC) interval. Resistance to extinction was significantly greater following 150 as compared to 250 acquisition trials. In addition, the 2-sec CS–UCS interval resulted in slower acquisition and slightly less rapid extinction than the 4-sec interval. Conditioned planaria also showed significant spontaneous recovery from the first to the second day of excitation. It was concluded that optimal numbers of CS–UCS pairings exist for planaria as for the other classically conditionable organisms.

Koopowitz, H., 1970, Feeding behavior and the role of the brain in the polyclad flatworm, *Planocera gilchristi, Anim. Behav., 18,* 31–34.

The feeding behavior of *Planocera* consists of a complex sequence of events. They feed by capturing snails and extracting the molluscs from their shells. Decerebration of the flatworms affects both the sequential timing of the patterns and the ability to capture prey, but brainless animals are still capable of feeding.

Koopowitz, H., 1972, Organization and physiology of flatworm nervous systems, *Am. Zool., 10,* 549–550.

Morton, R. and Kleinginna, P., 1971. Running speed and mortality rate of *Bipalium kewense* as a function of different levels of illumination, *J. Biol. Psychol.,* 13, 25–26.

Planarians, *Bipalium kewense*, were found to follow the grain in a straight wooden runway. Running speed was directly proportional to the body length. The intensity of an overhead incandescent light was varied and found to be related to mortality rate. An escape paradigm led to no signs of learning, and running speed was not significantly related to illumination level.

Behavior Modification in Annelids (Vol. 1)

Doolittle, J. H., 1971, The effect of thigmotaxis on negative phototaxis in the earthworm, *Psychonomic Sci. Sect. Anim. Physiol. Psychol.*, *22*, 311–312.

Negative phototaxic (light avoidance) responses of earthworms were measured under two thigmotaxic (bodily contact) conditions: presence or absence of an alley. As predicted by Smith (1902), the presence of an alley greatly increased the time taken by Ss to move away from a light source. It was concluded that thigmotaxis is a powerful motivational variable in the earthworm.

Ratner, S. C., 1972, Habituation and retention of habituation in leech, *Macrobdella decora*, *J. Comp. Physiol. Psychol.*, *81*, 115–121.

Ratner, S. C., and Gelpin, A. R., 1974, Habituation and retention of responses to air puff in normal and decerebrate earthworms, *J. Comp. Physiol. Psychol.*, *86*, 911–918.

Ward, J. E., and Doolittle, J. H., 1973, The effect of the anterior ganglia on forward movements in the earthworm, *Physiol. Psychol.*, *1*, 129–132.

The role of the anterior ganglia in the ability of the earthworm (*Lumbricus rubellus*) to inhibit movement when substrate contact was lost and to move forward was investigated. In Experiment I, worms were observed during the regeneration period following removal of the anterior five segments. A difference ($p = .005$) was found between the experimental (removal) and control (nonremoval) groups in ability to prevent falling off an elevated platform. The proportion of forward movements for the experimental group increased from an initial low of 0.31 to 0.93 over 31 days, while the control group maintained a high proportion (0.97) throughout the experiment. Experiment II replicated the first 36-hr period of Experiment I and confirmed the finding that the proportion of forward movements for the experimental group declined slowly to a minimum at 24 hr, then increased to 0.21 by 36 hr. The results were discussed in relation to Needham's (1957) hypothesis of a posterior hormonal and metabolic integrating center.

The Chelicerates (Vol. 2)

Adolph, A. R., and Tuan, F. J., 1972, Serotonin and inhibition in *Limulus* lateral eye, *Gen. Physiol.*, *60*, 679–697.

Bursey, C. R., 1973, Microanatomy of the ventral cord ganglia of the horseshoe crab, *Limulus polyphemus* (L.), *Zellforsch. mikrosk. Anat.* *137*, 313–329.

Cloudsley-Thompson, J. L., 1973, Entrainment of the "circadian clock" in *Buthotus minax* (Scorpiones: Buthidae), *J. Interdiscipl. Cycle Res.*, *4*, 119–123.

Fein, A., and De Voe, R. D., 1973, Adaptation in the ventral eye of *Limulus* is functionally independent of the photochemical cycle, membrane potential, and membrane resistance, *J. Gen. Physiol.*, *61*, 273–289.

Graham, N., Ratliff, F., and Hartline, H. K., 1973, Facilitation of inhibition in the compound lateral eye of *Limulus, Proc. Nat. Acad. Sci.* U.S.A., *70*, 894–898.

Herman, W. S., 1970, *Limulus* neuroendocrinology, *Am. Zool.*, *10*, 497.

Lahue, R., and Corning, W., 1973, Incremental and decremental processes in *Limulus* ganglia: Stimulus frequency and ganglion organization influences, *Behav. Biol.*, *8*, 637–654.

Lahue, R. H., 1974, Habituation characteristics and mechanisms in the abdominal ganglia of *Limulus polyphemus*, doctoral dissertation, University of Waterloo, Waterloo, Ontario, Canada.

Szlep, R., 1964, Change in the response of spiders to repeated web vibrations, *Behaviour, 23,* 203–238.

Repeated rhythmic recurrence of web vibration brings about a waning of response not due to motor fatigue. Extinction was not caused by adaptation of the receptors to a given vibration intensity. Delayed actions and incomplete responses suggest the appearance of inhibiting impulses. Rise of threshold level causes weakening and slowing down of the response, while the appearance of inhibiting impulses brings about waning of the response to vibrations.

Learning in Crustacea (Vol. 2)

Atwood, H. L., and Lang, F., 1972, Synaptic vesicles: Selective depletion in crayfish excitatory and inhibitory axons, *Science* Wash., D.C., 176, 1353–1355.

Berrill, M., 1971, The embryonic development of the avoidance reflex of *Neomysis americana* and *Praunus flexuosus* (Crustacea: Mysidacea), *Anim. Behav.*, *19*, 707–713.

The embryonic antecedents of the avoidance reflex of both *Praunus flexuosus* and *Neomysis americana* appear to be spontaneous before they become reflexive. The spontaneous flinches of the naupliar embryo appear to become the twitches of the postnaupliar embryo, which in turn become the avoidance reflexes of the juvenile. The postnaupliar twitches are also apparently spontaneous at first, but they occur increasingly often in response to tactile stimulation as postnaupliar development progresses. Though each embryonic molt results in a sudden emergence of greater complexity of embryonic activity, there is gradual elaboration of the activity between such molts.

Bruner, J., and Kehoe, J., 1970, Long-term decrements in the efficacy of synaptic transmission in molluscs and crustaceans, *in* "Short-Term Changes in Neural Activity and Behaviour" (G. Horn and R. A. Hinde, eds.), pp. 323–359, Cambridge University Press, Cambridge, England.

Chapple, W. D., 1973, Changes in abdominal motoneuron frequency correlated with changes of shell position in the hermit crab *Pagurus pollicarus*, *J. Comp. Physiol.*, *87*, 49–64.

Chow, K., and Leiman, A. L., 1972, The photo-sensitive organs of crayfish and brightness learning, *Behav. Biol.*, *7*, 25–35.

The differential properties of the compound eye system and caudal photoreceptors of crayfish in the mediation of simple, light-controlled behaviors were examined in habituation and avoidance learning experiments. The initial response of crayfish to a bright light consists of a rapid tail flick, which produces backward motion. This escape response rapidly habituates and is dependent on compound eye input, i.e., eyeless animals do not exhibit this behavior, although visual input is available through the operation of caudal photoreceptors. A slower, withdrawal response involving locomotion also readily habituates during the course of regularly repeated light stimuli. However, this decrement in responsiveness is independent of the source of light input, since it was demonstrated in groups of normal animals, eyeless animals, and crayfish without functional caudal photoreceptors. Conditioned avoidance response learning was also demonstrated in all groups of animals. Some of these data suggest the establishment of two parallel neural systems mediating this form of learning, one connected to the eyes, the other to the tail photoreceptor.

Costanzo, D. J., and Cox, W. G., 1971, Habit reversal improvement in crayfish, *J. Biol. Psychol.*, *13*, 11–12.

Two groups of male and female crayfish each learned a combined visual-spatial discrimination with contingencies of reward alone or reward plus punishment. Once the discrimination was learned, the crayfish were taught nine daily reversals and demonstrated overall improvement with successive reversals. Analysis of sex and reinforcement conditions revealed no differences or interactions in initial learning or habit reversal improvement.

Costanzo, D. J., Rudolph, G. R., and Cox, W., 1972, Social status and habit reversal learning in crayfish, *J. Biol. Psychol.*, *14*, 30–32.

Groups of dominant and nondominant male crayfish were trained on a combined visual-spatial discrimination problem. Fifteen massed trials were given for training, reversal, and extinction. An analysis of variance showed a significant difference between dominant and nondominant groups during reversal learning, but not for initial training or extinction. However, there was an indication that the dominant Ss were persistent in their level of responding during extinction trials.

Coward, S. J., Gerhardt, H. C., and Crockett, D. T., 1970, Behavioral variation in natural populations of two species of fiddler crabs (*Uca*) and some preliminary observations on directed modifications, *J. Biol. Psychol.*, *12*, 24–31.

Fay, R. R., 1973, Multisensory interaction in control of eye-stalk rotation response in the crayfish (*Procambarus clarkii*), *J. Comp. Physiol. Psychol.*, *84*, 527–533.

Glantz, R. M., 1971, Peripheral versus central adaptation in the crustacean visual system, *J. Neurobiol., 34,* 485–492.

Response adaptation, the increment threshold function, and dark adaptation were studied in peripheral retinula cells and central sustaining fibers. A comparison of results at the two levels indicates that response adaptation occurs more rapidly and to a far greater degree in the central nervous system than in the receptor. Thus the sustaining fiber steady-state discharge seldom exceeds 3% of the maximal firing rate (300 per sec). This low rate precludes a saturation-dependent loss of gain as observed in the photoreceptor. As a consequence the desensitization associated with the increment threshold function and dark adaptation is confined to the periphery. When these functions are monitored in optic nerve interneurons the thresholds merely reflect the adaptive state of the retinula.

Glantz, R. M., 1973. Five classes of visual interneurons in the optic nerve of the hermit crab, *J. Neurobiol., 4,* 301–319.

Greene, R. L., and Jennings, J. W., 1970, Apparati for assessing maze learning in the isopod (*Porcellio scaber*), *J. Biol. Psychol., 12,* 37–40.

This study was designed to test the differences between some typical maze cross sections and mazes to define an apparatus adequate for testing maze learning in isopods. The study consisted of two parts. Straight-alley latency data were recorded for square, semicircle, circle, and triangle cross sections. Ss ran significantly faster in a square cross-section alley and significantly slower in a triangular cross-section alley, with circle and semicircle cross sections intermediate. Then escape responses were recorded for T- Y- and cross mazes. Learning occurred in all apparati. The T- and cross mazes produced significantly better performance than the Y-maze.

Harless, M., 1971, Isopods as subjects for behavioral research, *J. Biol. Psychol., 13,* 13–15.

Krasne, F. B., and Bryan, J. S., 1973, Habituation: Regulation through presynaptic inhibition, *Science* Wash., D.C., *182,* 590–592.

During tail-flip escape responses of crayfish, synaptic transmission at the habituation-prone synapses of the lateral giant reflex pathway is presynaptically inhibited. This prevents transmitter release and all subsequent postsynaptic actions and spares the reflex from becoming habituated to stimuli produced by an animal's own escape movements. These observations demonstrate the existence of a control circuit whose adaptive function is to regulate the malleability of inherently plastic synapses. They also suggest that regulation of plasticity could be a common use of presynaptic inhibition.

Millechia, R., and William, G. F., 1972, Photoreception in a barnacle: Electrophysiology of the shadow reflex pathway in *Balanus cariosus*, *Science* (Wash., D.C.), *177,* 438–441.

Mittenthal, J. E., and Wine, J. J., 1973, Connectivity patterns of crayfish giant interneurons: Visualization of synaptic regions with cobalt dye, *Science* (Wash., D.C.), *179,* 182–184.

Nakajima, Y., Tisdale, A. D., and Henkart, M. P., 1973, Presynaptic inhibition at inhibitory nerve terminals. A new synapse in the crayfish stretch receptor, *Proc. Nat. Acad. Sci. U.S.A., 70,* 2462–2466.

Page, C. H., and Sokolove, P. G., 1972, Crayfish muscle receptor organ: Role in regulation of postural flexion, *Science* (Wash., D.C.), *175*, 647–650.

Paul, D. H., 1972, Decremental conduction over "giant" afferent processes in an arthropod, *Science* (Wash., D.C.), *176*, 680–682.

Smith, D. O., 1972, Central nervous control of presynaptic inhibition in the crayfish claw, *J. Neurophysiol.*, *35*, 333–343.

Taylor, R. C., 1971, Instrumental conditioning and avoidance behavior in crayfish, *J. Biol. Psychol.*, *13*, 36–41.

Crayfish (S.) show an age-dependent light-avoidance response that exhibits an initial decrease and then returns to an intermediate level by the sixtieth trial. Ss learned not to respond when shocked every time they pass through the dark chamber of a shuttlebox in response to a light. Ss learned to shuttle in response to a single light in the middle of the shuttlebox. The learning rate was most rapid with a 1-sec CS–UCS interval and no learning was shown with a 10-sec CS–UCS interval. Extinction curves for both types of learning were very rapid.

Walker, I., 1972, Habituation to disturbance in the fiddler crab (*Uca annulipes*) in its natural environment, *Anim. Behav.*, *20*, 139–146.

Crabs of both sexes and of various sizes were subjected to a standardized test of ten consecutive disturbances in their natural habitat. Males and females reacted alike: 53% became fully habituated within less than ten disturbances, and a further 12% showed slower habituation. The remainder of the test individuals showed various other patterns of reactions. Reaction pattern and efficiency of habituation are size-dependent (or age-dependent) and they are modifiable by social interaction and by tide-determined activity phases. The structure and species composition (with regard to ocipodid crabs only) of the *Uca* habitat is considered briefly, and some observations on display and "rapping" or drumming behavior are reported.

Wiersma, C. A. G., and Yanagisawa, K., 1971, On types of interneurons responding to visual stimulation present in the optic nerve of the rock lobster, *Panulirus interruptus*, *J. Neurobiol.*, *2*, 291–309.

York, B., 1972, Sustaining fibres in the rock lobsters, *J. Neurobiol.*, *3*, 303–309.

Learning in Insects and Apoidea (Vol. 2)

Beale, I. L., and Webster, D. M., 1971, The relevance of leg movement cues to turn alternation in woodlice (*Porcellio scaber*), *Anim. Behav.*, *19*, 353–356.

When wood lice were passively rotated through 90°, rather than making the usual forced active turn, they did not show turn alternation. Also, passive rotation through 90° in the same or opposite direction as a prior forced active turn did not affect the likelihood of subsequent alternation relative to the original turn. When a moving track on one side of a runway caused the legs on the corresponding side of a passing wood louse to move more rapidly than its other set of legs, it was subsequently more likely to turn to the side on which its legs had been more active. Taken together, these results suggest that cues arising from differential leg movements during a forced turn may be critical to turn alternation.

Bell, W. J., Burk, T., and Sams, G. R., 1973, Cockroach aggregation pheromone: Directional orientation, *Behav. Biol., 9,* 251–255.

Orientation of adult and immature cockroaches, *Periplaneta americana,* in a T-maze is directed by a feces-secreted aggregation pheromone. Tropochemotactic orientation is dependent on olfactory perception of the pheromone by antennal receptors.

Bennet-Clark, H. C., Ewing, A. W., and Manning, A., 1973, The persistance of courtship stimulation in *Drosophila melanogaster, Behav. Biol., 8,* 763–769.

Five minutes of prestimulation with simulated courting sound before mixing with males leads to enhanced receptivity of female *Drosophila melanogaster.* The effects of prestimulation fade and can no longer be detected after 5 min. One minute of sound was ineffective and so were sounds with pulse intervals one-half and twice the natural interval. It is concluded that the female not only can summate stimuli over time but that the effects of stimuli perseverate for some minutes in the absence of continuing stimulation.

Bentley, D. R., and Hoy, R. R., 1972, Genetic control of the neuronal network generates cricket (*Teleogryllus, Gryllus*) song patterns, *Anim. Behav., 20,* 478–492.

Hybrid field crickets between *Teleogryllus* species and *Gryllus* species were produced to examine: (1) the genetic controlling the neuronal network underlying song production, and (2) the neuronal mechanism responsible for the superimposed rhythms of chirping songs. Pulse numbers, intervals, and progressions were subjected to statistical and graphic analyses. *Teleogryllus* songs are controlled by a complex, polygenic, multichromosomal system, even at the level of "unitary" acoustical parameters. Genes regulating certain pattern characteristics are on the X-chromosome. The superficially similar chirping songs of two *Gryllus* species appear due to different neuronal mechanisms. One song appears to reflect a single rhythm generator with some relaxation oscillation properties, while the rhythms of the other song seem to be caused by the mixed effect of two resonant-type oscillators.

Bermant, G., McNeil, M., and Ashby, F. N., 1973, Honey bee behavior: Responses to light and electric shock, *Behav. Biol., 9,* 505–509.

The position of worker bees in a shuttlebox was controlled by variation of the intensities of light stimuli placed at both ends of the box. The amount of time spent on the brighter side of the box was greatest when the gradient of illumination was steepest. Responding to light was altered systematically by the delivery of electric shock of the legs. We suggest that electric shock may be used to calibrate the strength of phototactic responding.

Blake, G. M., 1970, An imcomplete randomized block design, illustrated by a study of humidity discrimination in *Anthrenus verbasci., Anim. Behav., 18,* 96–102.

The humidity preferences of both adults and young larvae of *Anthrenus verbasci* have been studied in an alternative chamber. The results were analyzed by means of an incomplete randomized block design. The humidities, presented in all possible pairs in the chamber, were 30%, 50%, 70%, and 90% relative humidity. The insects used in all four experiments were able to discriminate among the four humidity levels and generally congregated in the lower of any pair presented. The power of discrimination between any two humidities was proportional to the difference between them and was not dependent upon the actual values of the humidities involved.

Burrows, M., and Hoyle, G., 1972, Neural mechanisms underlying behavior in the locust *Schistocerca gregaria.* III. Topography of limb motorneurons in the metathoracic ganglion, *J. Neurobiol., 4,* 167–186.

Campbell, S. L., and Holgate, S. H., 1971, Temperature selection in two subspecies of beetle, *J. Biol. Psych., 13,* 16–17.

Temperature selection frequencies in a thermal gradient was recorded for four members of each of two subspecies of beetle from the same geographical region. While the two subspecies did not differ significantly in temperatures selected, they did differ with respect to distribution of selection frequencies and number of minute-to-minute position changes. The latter difference was significant at the 0.01 level. The difference was ascribed to an interaction between differences in running speeds of the two subspecies and temperature changes in the cercal nerves of the sixth abdominal ganglion.

Carlson, A. D., and Copeland, J., 1972, Photic inhibition of brain stimulated firefly flashes, *Am. Zool. 12,* 479–487.

Chen, W. Y., Aranda, L. C., and Luco, J. V., 1970, Learning and long- and short-term memory in cockroaches, *Anim. Behav., 18,* 725–732.

Intact and headless roaches were trained to avoid electric shock when an electric signal was applied just before shock delivery. Intact animals learned the response and showed no decrement during the experiment; removing the head or a ganglion after learning did not affect retention. Headless or isolated ganglion preparations did not retain as long (2–3 days and 1 day, respectively).

Dagan, D., and Purnas, I., 1970, Giant fibre and small fibre pathways involved in the evasive response of the cockroach, *Periplaneta americana, J. Exp. Biol., 52,* 313–324.

Giant fibers were found not to activate leg motoneurons during evasion. A pathway of small axons having a conduction velocity of 1.5–4.5 m per sec was found to govern leg activation during escape. This pathway remains functional after giant-fiber degeneration after the giant fibers have been severed from their somata. Movements of the antennae were found to be activated by the giant fibers simultaneously with or slightly earlier than movements of the legs. This suggested that a general alarm system is activated by the giant fibers concomitantly with activation of the leg motoneurons by a slower conduction pathway.

Dobrzanski, S., 1971, Manipulatory learning in ants, *Acta Neurobiol. Exp.* (Warsaw), *31,* 111–140.

The building behavior of ants consists of hereditary and acquired elements. Hereditary reactions are seizing building materials, transporting them on the summit of the mound, and putting them down parallel to the surface. Acquired reactions are the skill of selecting building materials suitable for transportation, skill at carrying materials, overcoming obstacles, and mounting material into the mound structure. The presence of manipulatory learning in ants alters views concerning their individual differences and the plasticity of their behavior.

Fukushi, T., 1973, Olfactory conditioning in the housefly, *Musca domestica, Annot. Zool. Jap., 46,* 135–143.

Houseflies, *Musca domestica* Linne, were conditioned to the odors of alcohols, n-butyraldehyde, and acetic acid by using proboscis extension to labullar stimulation with sucrose solution as an unconditioned response. No difference was found between males and females in conditionability. The flies could discriminate among the three odors of the conditioned stimuli, but they were unable to do so among the members of a lower alcohol series containing ethanol, *n*-propanol, isopropanol, *n*-butanol, and tert-butanol. The conditioned responses of the flies were inhibited by addition of a new odor to the one originally used in conditioning. Flies conditioned to a particular odor could, however, be conditioned to second and third odors independently. Heads of conditioned flies that had been severed from their bodies continued to respond to the conditioned stimulus. Experimental extinction was also observed. The olfactory receptors of the housefly seem to be located mainly at the basal segment of the antennae.

Grossmann, K. E. von, 1970, Erlernen von Färbreizen an der Futterquelle durch Honigbienen während des Anflugs und während des Säugens (Learning of color stimuli at a feeding site by honeybees during approach, landing, and sucking), *Z. Tierpsychol., 27,* 553–562.

Honeybees landed on a transparent cone that was illuminated from beneath with orange light. When they sucked a 40% sugar solution, the light changed from orange to blue. During choice tests, the bees landed on orange cones 75% of the time and approached orange 60% of the time. Control bees landed and sucked on orange cones. During tests, they landed on orange 90% and approached orange 80% of the time. In a second experiment, bees landed on yellow pieces of cardboard. After landing, they were carried passively to a blue one. During tests, they landed on yellow 67% of the time and approached yellow 60% of the time. Control bees landed and sucked on yellow. During tests they landed on yellow 100% of the time and approached yellow 90% of the time. The results indicate that if the color during the sucking phase is different from the color during landing at a feeding place, bees learn both colors. The color that is present during landing, however, dominates even if there is a delay of 3–10 sec between landing and reinforcement. The ability of honeybees to learn color signals even when reinforcement is delayed might possibly be understood in terms of innate behavior patterns adapted for nectar collection. Behavioral analysis may be possible if two operants are assumed: landing and preparatory sucking activities.

Grossmann, K. E. von, 1971, Belohnungsverzögerung beim Erlernen einer Farbe an einer künstlichen Futterstelle durch Honigbienen (Color learning under conditions of immediate and delayed reinforcement by honeybees), *Z. Tierpsychol., 29,* 28–41.

When naïve bees are recruited by experienced foraging bees, they learn a specific scent that enables them to find proper feeding places with increased probability. However, this scent is communicated by the foragers as a secondary reinforcement, which perhaps exists without any direct association with a primary reinforcer. Color learning may be facilitated in this manner.

Grossmann, K. E. von, and Beller, E. J., 1971, Das Erlernen unbelohnter Farbsignale durch Honigbienen (Discrimination learning in honeybees: The avoidance of unreinforced hues), *Z. Tierpsychol.*, *29*, 449–446.

1. Honeybees were trained to land on a plexiglass disc 20 cm in diameter. They received a drop of syrup 8 mm below a 5-mm round opening in the center of the disc upon interruption of an infrared light beam. Subsequently they were rewarded only in the presence of S^+ (504 nm) and not rewarded in the presence of S– (489 nm or 535 nm, respectively). These hues were projected onto a frosted glass disc 6 cm beneath the plexiglass disc. 2. After 20 visits with alternating presentation of S^+ and one of the two Ss, the bees entered the hole more frequently in the presence of the other, unknown S^0 during the twenty-first visit. Upon alternating presentation of both S^- and S^0, the bees entered the hole more often during the S^- intervals. 3. When the bees were prevented from entering the central opening during the S^- intervals they still learned to prefer S^0 over S^- 4. When, in addition, landing was also prevented during either S^+ or S^- intervals, the bees still prefered S^0 over S^-. 5. When even crawling behavior was prevented by means of a short period of confinement, the bees also acted less frequently during the S^- intervals than during the S^0 intervals; the difference, however, was not as prominent as with free discrimination behavior. 6. The preference of S^0 to S^- was learned later than was the preference of S^+. After one visit the S^0 intervals were chosen 70% of the time, after ten visits 90%. 7. The learning process was not restricted to unrewarded colors in the range between 489 and 535 nm but was also demonstrated for 430 and 170 nm.

Hoyle, G., and Burrows, M., 1973*a*, Neural mechanisms underlying behavior in the locust *Schistocerca gregaria.* I. Physiology of identified motoneurons in the metathoracic ganglion, *J. Neurobiol.*, *4*, 3–41.

Hoyle, G., and Burrows, M., 1973*b*, Neural mechanisms underlying behavior in the locust *Schistocerca gregaria.* II. Integrative activity in metathoracic neurons. *J. Neurobiol. 4*, 43–67.

Jander, R., 1971, Visual pattern recognition and directional orientation in insects, *Ann. N. Y. Acad. Sci.*, *188*, 5–11.

Jander, R. von, Fabritus, M., and Fabritus, M., 1970, The significance of disruption and contour-direction for visual form discrimination near the nest entrance in the wasp *Dolichovespula saxonica*, *Z. Tierpsychol.*, *27*, 881–893.

Closed, filled-in shapes are preferred to disrupted patterns at the nest entrance. Animals can be trained to prefer disrupted (checkerboard) patterns. Discrimination between bars of different tilt angles can also be acquired if there is a straight bar for reference.

Kafka, W. A., Ohloff, G., Schenider, D., and Vareschi, E., 1973, Olfactory discrimination of two enantiomers of 4-methylhexanoic acid by the migratory locust and the honeybee, *J. Comp. Physiol.*, *87*, 277–284.

Single olfactory receptor cells of the migratory locust and of honeybee drones were stimulated with the two enantiomeric forms of 4-methylhexanoic acid. In the majority of cells, equal stimulus quantities of either the (+) or the (–) isomer elicited a greater number of nerve impulses. Behavioral discrimination of the two enantiomers was demonstrated in drones. Here, the proboscis extension response served as indicator of the discrimination following the condition to one or the other steric antipode. These findings are explained by a minimum of two variants of one structurally highly stabilized type of acceptor present in different numbers on the membranes of individual cells of this type.

Kerkut, G. A., Emson, P. C., and Beesley, P. W., 1972, Effect of leg-raising learning on protein synthesis and ChE activity in the cockroach CNS, *Comp. Biochem. Physiol.*, *41B*, 635–645.

The normal electrophoretic pattern for cockroach-ganglia–soluble proteins was established. At least 16 protein fractions were detectable. The band pattern was consistent and reproducible. A double labeling technique indicated that 3 protein fractions showed increased incorporation of labeled amino acid during learning. Cholinesterase activity in the metathoracic ganglion extracts was investigated by disc electrophoresis followed by spectrophotometric analysis. The supernatant of 10,000 g for 1 hr revealed the presence of two major isoenzymes. The activity of the ChE isoenzymes was studied after learning. There was a decrease in the activity of the major cholinesterase isoenzyme. This was due to a change in the K_m of the enzyme from $5.88 \mathrm{X} 10^{-5}$M in the yoke controls, to 1.33×10^{-4}M in the trained animals.

Koltermann, R., 1971, 24-Std-Periodik in der Langzeiterinnerung an Duft- und Farbisignale bie der Honigbiene (Circadian memory rhythm after scent and color training with honeybees), *Z. Vgl. Physiol.*, *75*, 49–68.

Bees can learn to associate a scent with the time of day that it was learned. Two such associations can be learned with intervals of 6, 3, and 2 hr and 45 min. Two different scents are learned at a different time, providing the scents are equal in their attractiveness. Colors can also be associated with time in a 24-hour period.

Koltermann, R., 1973, Retroaktive Hemmung nach sukzessiver Informationseingable bei *Apis mellifica* und *Apis cerana* (Apidae) (Retractive memory inhibition in *Apis mellifica* and *Apis cerana* after learning two kinds of information in succession), *J. Comp. Physiol.*, *84*, 299–310.

When two scents are learned successively in a feeding response, the second food signal interferes with response to the first. Comparisons between signals of different attractiveness were made.

Kriston, I., 1971, Zum Problem des Lernverhaltens von *Apis mellifica L.* gegenüber verschiedenen Duftstoffen (To the problem of the learning behavior of *Apis mellifica L.* to different odorous substances), *Z. Vgl. Physiol.*, *74*, 169–189.

Although there were no spontaneous preferences demonstrated for nine different scents and one flower scent, learning efficiency varied according to the scent.

Leake, L. D., and Taylor, I. B., 1972, The role of GABA in learning in the cockroach? *Comp. Gen. Pharmacol.*, *3*, 469–472.

Injection of NH_2OH (hydroxylamine) and AOAA (amino-oxyacetate) into headless cockroaches decreased learning time. INH (isonicotinic acid hydrazide) increased learning time, while semicarbide had no effect. It is inferred that learning is accompanied by an increase in GABA levels. A model, based on this assumption, is proposed to account for the possible role of GABA in the learning process. It postulates simultaneous facilitation of the specified (learning) pathway by ACh and inhibition of alternative pathways by GABA.

Loher, W., and Chandrashekaran, M. K., 1972, Communicative behavior of the grasshopper *Syrbula fuscovittata* with particular consideration of the male courtship, *Z. Tierpsychol.*, *31*, 78–97.

Courtship behavior can be modified by external cues, although the structure of specific elements of the courtship song remains unchanged.

Lovell, K. L., and Eisenstein, E. M., 1973, Dark avoidance learning and memory disruption by carbon dioxide in cockroaches, *Physiol. Behav.*, *10*, 835–840.

Cockroaches were trained to avoid the dark side of a box with electric shock used as a negative reinforcement. The majority of the acquisition of learning occurred within the first minute of training. The reduction in the number of shock bouts initiated during training could be attributed entirely to avoidance learning; no evidence of escape learning was found. There was no decrease in retention of dark avoidance up to 2 hr after training. When carbon dioxide was administered immediately after training, no retention was observed 2 hr later. When CO_2 was given 1 hr after training, some retention was observed 2 hr after training. These results indicate that memory phases that have different susceptibilities to disruption can be observed in cockroaches.

Luco, J. V., 1971, A study of memory in insects, *in* "Research in Physiology" (F. F. Kao, K. Koizumi and M. Vasalle, eds.), Aulo Gaggi, Bologna.

Maldonado, H., 1972, A learning process in the praying mantis, *Physiol. Behav.*, *9*, 435–445.

Adult female mantids (*Stagmatoptera biocellata*) were shown a star or a fly, but were prevented from catching it. After several frustrated attacks there was a clear-cut decrease of the mantises' responses. The fade-out became apparent through two manifestations, i.e., a fall in the number of strikes and a reduction in the amplitude of the head movements. The persistence of the decrement proved to last at least eight days, and an analysis of the parametric characteristics of the waning response allowed us to define it as a learning process. The category of learning to which this process should be allocated was discussed. To obtain more information about this learned behavior an analysis of the relationship between attacks and head movements was done, i.e., between both symptoms of the process.

Murphey, R. K., 1973, Mutual inhibition and the organization of a non-visual orientation in *Notonecta*, *Comp. Physiol.* *84*, 31–40.

Hypotheses concerning the neural mechanisms by which the back swimmer *Notonecta undulata* locates prey have been examined by the use of behavioral tests. The results are consistent with the following hypotheses: (1) the receptor nearest the target controls the direction of the turn, which is elicted by a stimulus; (2) sensory input via a given receptor is capable of eliciting a very limited range of motor responses; and (3) there are inhibitory interactions between receptors at some level of the central nervous system. A neuronal network analogous to a lateral inhibitory network is proposed to be the neural basis for the orientation.

Oliver, G. W. O., Taberner, P. V., Rick, J. T., and Kerkut , G. A., 1971, Changes in GABA level, GAD and ChE activity in CNS of an insect during learning, *Comp. Biochem. Physiol. 38B*, 529–535.

Cockroches that learn to keep their legs lifted in the air (experimental animals) have (1) lower GAD activity in their thoracic ganglia; (2) lower levels of GABA in their thoracic ganglia; (3) a lower GABA/glutamate ratio in their head; and (4) lower ChE activity in their head. The experimental animals were compared with control animals (equal number of shocks but not learning) and resting animals (no shocks). In all cases the level of GAD, GABA, GABA/glutamate, and ChE was resting animals $>$ control animals $>$ experimental animals.

Payne, J. H., 1972, Cockroaches, *Periplaneta americana:* A preliminary frigid study, *J. Biol. Psychol., 14*, 36–37.

Temperatures between 50°F and 35°F were studied to determine their effect upon retention in cockroaches having learned to avoid the dark end of a T-maze. Ss were exposed to temperature conditions for 1 hr and given 1-hr recovery before retention trials. Cold-temperature exposure was found to inhibit retention significantly; this inhibition was found to be only temporary. There may be behavioral evidence that this may be temporary neural inhibition.

Pinel, J. P. J., and Mucha, R. F., 1973, Incubation effect in the cockroach (*Periplaneta americana*), *J. Comp. Physiol. Psychol., 85*, 132–138.

In Expt. 1, activity of cockroaches was recorded in an open field in which they had received footshock (FS) 1, 3, 10, 30, or 120 min before. The FS produced an incubation effect, i.e., a general decrease in activity, with the greatest decrease occurring at the longer intervals. In Expt. 2, activity of cockroaches was recorded in an open field 120 min after one FS and 1, 3, 10, or 30 min. after a second. When no second FS was administered, the cockroaches were inactive; however, when testing occurred within 10 min of the second FS, activity was greatly increased. This gradient of hyperactivity paralleled exactly the incubation effect demonstrated in the first experiment. The results of both experiments are almost identical to the results of comparable experiments in which rats have served as subjects. Thus, the incubation effect is an adaptive behavior pattern resulting from the transient activating effects of noxious stimulation.

Pritchatt, D., 1970, Further studies on the avoidance behavior of *Periplaneta americana* to electric shock, *Anim. Behav., 18*, 485–492.

Rick, J. T., Oliver, G. W. O., and Kerkut, G. A., 1972, Acquisition, extinction and reacquisition of a conditioned response in the cockroaches: The effects of orotic acid, *Quart. J. Exp. Psychol., 24*, 282–286.

Cockroaches can be trained to keep their metathoracic leg out of a saline solution. Isolated metathoracic ganglia learned faster than headless animals, which in turn learned faster than intact animals. Extinction took longer in the isolated ganglion than in the other two preparations. Extinction times increased with increasing dose of orotic acid, a precursor of RNA. Orotic acid did not systematically affect the times for acquisition and reacquisition of the learned response.

Rockwell, R. F., and Seiger, M. B., 1973, Phototaxis in *Drosophila:* A critical evaluation, *Am. Sci., 61,* 339–345.

Schencking, M. S. von, 1970, Investigation of visual learning rate and learning capacity in bees, bumble-bees and ants, *Z. Tierpsychol., 27,* 513–551.

Stratton, L. O., and Coleman, W. P., 1973, Maze learning and orientation in the fire ant (*Solenopsis saevissima*), *J. Comp. Physiol., 83,* 1–12.

Acquisition and retention of location of a food source was studied in the fire ant by the use of a group learning criterion. Chemical, distal-visual, and kinesthetic cue changes led to increase in errors to criterion, but adjustment to changes in a single sensory cue was rapid once the workers had learned the maze. Although chemical cues are prepotent for the fire ant, learning was possible without them. Group learning curves were comparable to those of the single *Formica* ant studied by Schneirla.

Swihart, C. A., 1971, Colour discrimination by the butterfly, *Heliconius charitonius Linn., Anim. Behav., 19,* 156–164.

The color-discriminating ability of *Heliconius charitonius* was studied as they were conditioned to variously colored paper model flowers. These experiments demonstrated their ability to generalize between two shades of yellow and simultaneously revealed that the butterflies appreciated the differences. Whenever the training color was offered in the test array, that color would receive the greatest number of visits. Experiments were conducted to determine the butterflies' ability to distinguish between similar hues throughout the spectrum and gave no firm indication that certain colors could be discriminated more accurately than other hues. It appears that the butterflies can form a searching image for flowers that consist of several different hues. It was not possible to condition them to select a particular shade of gray.

Swihart, C., and Swihart, S., 1970, Colour selection and learned feeding preferences in the butterfly, *Heliconius charitonius Linn., Anim. Behav., 18,* 60–64.

Conditioning experiments involved feeding the animals on a single-color model and testing the feeding responses with the multicolored array. Of the feeding responses in the untrained animals, 9.1% were on the yellow flower and 2.0% on the green. When they were fed on yellow models, the number of visits on that color rose to 49.6%. When green models were employed for feeding, the responses to the model constituted 55.3% of the visits. A total of 1800 feeding attempts was recorded.

Taylor, R. J., 1974, Role of learning in insect parasitism, *Ecol. Monogr. 44,* 89–104.

Several stochastic models of the process of parasitism, each incorporating learning in a different way, are developed. The clues the parasitoid learns to find its host are divided into two types: clues to individual hosts and clues to the habitat can be a reflection of the spatial distribution of the hosts. The conclusion suggested by the experimental test of the models with the ichneumonid wasp *Nemeritis canescens* are that *Nemeritis* can learn to hunt its host in a novel environment; that a model postulating the learning of one; and that learning is potentially a stabilizing factor in the dynamics of host–parasite systems.

Wehner, R., and Toggweiler, F., 1972, Verhaltens-physiologischer Nachweis des Farbensehens bei *Cataglyphis bicolor* (Formicidae, Hymenoptera), *J. Comp. Physiol.*, *77*, 239–255.

The color vision of outside workers of desert ants was investigated by the training of individuals to respond to monochromatic lights. The spectral sensitivity distribution is characterized by a prominent peak in the ultraviolet region and two less prominent peaks. Ants chose wavelengths to which they had been trained when offered a choice of wavelengths. A function for determining the precision with which wavelengths can be discriminated by ants is suggested. This function has two maxima, which coincide with the minima of the three-peaked spectral sensitivity distribution, offering evidence for a trichromatic color system.

Willner, P., and Mellanby, J., 1974, Cholinesterase activity in the cockroach CNS does not change with training, *Brain Res.*, *66*, 481–490.

Headless cockroaches were trained to hold a metathoracic leg in a raised position in order to avoid electric shock, and acetylcholinesterase (AChE) activity in the metathoracic ganglion was measured. The following factors were varied: age of the animals (nymph or adult), shock size and duration (0.8 mA, 2.0 msec, or 0.4 mA, 0.2 msec), method of immobilization (CO_2, anesthesia, or cooling), position of the shock electrodes on the preparation (both in the leg, or one in the leg and one in the neck), and the treatment following training (none, CO_2, anesthesia, or cooling). All AChE assays were performed both immediately after the ganglion had been homogenized, and again after 24 hr. No evidence could be found that AChE is altered by training.

Woodson, P. H. J., Schlapfer, W. T., and Barondes, S. H., 1972, Postural avoidance learning in the headless cockroach without detectable changes in ganglionic cholinesterase, *Brain Res.*, *37*, 348–352.

The report confirms the finding that the headless cockroach can learn a postural avoidance response of the metathoracic leg. Training was successful by four criteria. The failure to find changes in cholinesterase levels in the metathoracic ganglia of the trained animals, despite clear-cut evidence of learning suggests that the large changes in cholinesterase reported by Kerkut are not inextricably related to learning in this preparation. The authors give suggestions for the different results.

Yeatman, F. R., and Hirsch, J., 1971, Attempted replication of, and selective breeding for, instrumental conditioning of *Drosophila melanogaster*, *Anim. Behav.*, *19*, 454–462.

Our attempt to replicate fails to confirm previously reported instrumental conditioning of *Drosophila melanogaster.* Furthermore, ten generations of bidirectional selective breeding (eight generations by one criterion then two by another) have yielded no response to selection for apparent individual differences in performance on the conditioning test.

Yinon, V., Shulor, A., and Tsuilich, R., 1971, Audition in the desert locust: Behavioral and neurophysological studies, *J. Exp. Biol.*, *55*, 713–725.

Zilber-Gachelin, N. F., and Chartier, M. P., 1973*a*, Modification of the motor reflex responses due to repetition of the peripheral stimulus in the cockroach. I. Habituation at the level of an isolated abdominal ganglion, *J. Exp. Biol.*, *59*, 359–381.

In the cockroach repeated air puffs to the cerci induce a habituation of the corresponding escape reflex. Transmission through the first relay, the sixth abdominal ganglion was studied both by electrical stimulation of the sensory cercal nerve and by application of natural stimuli (air puffs) on the cerci, while recording simultaneously on the cercal nerve and the abdominal cord, i.e., respectively before and beyond the synaptic relay. The electrical stimulation study shows transmission to be relatively labile. The ganglionic relay is implicated as the center of a habituation phenomenon. The characteristics of this habituation are the following: with repetition of the stimuli, the response decreases down to a nonzero plateau; it recovers spontaneously if stimulations are suppressed; habituation is potentiated when successive series of habituations followed by spontaneous restorations are accumulated; it is more pronounced as the stimulation frequency is increased; finally, it does not show generalization.

Zilber-Gachelin, N. F., and Chartier, M. P., 1973*b*, Modification of the motor reflex responses due to repetition of the peripheral stimulus in the cockroach. II. Conditions of activation of the motoneurones, *J. Exp. Biol.*, *59*, 383–403.

The synaptic transfer properties within the third thoracic ganglion from the abdominal cord axons to the motoneurons has been studied in the cockroach. This ganglion was completely deafferented except for motor nerve 4, whose links with the muscles of the posterior legs were left intact. The response of one of these nerves to electrical stimulation of the abdominal cord involves activation of two types of units: slow excitatory fibers, which have a tonic discharge, and fast excitatory fibers, which have no tonic discharge. During repetition of the stimuli, the two types of synaptic pathways show both habituation and facilitation through temporal summation and posttetanic potentiation. These two phenomena persist after the end of the stimulations. Facilitation and habituation appear to be responsible for the sensitization of the responses that can be sometimes observed. They lead to the disappearance of the escape reflex involving firing of both fast and slow fibers.

Learning in Gastropod Mollusks (Vol. 2)

Berry, R. W., and Cohen, M. J., 1972, Synaptic stimulation of RNA metabolism in the giant neuron of *Aplysia Californica*, *J. Neurobiol.*, *3*, 209–222.

Black, D., Peretz, B., and Moller, R., 1972, Inhibitory influence of the CNS on habituation of the *Aplysia* gill withdrawal reflex, *Fed. Proc.*, *31*, 405.

A study of habituation of the *Aplysia* gill withdrawal response in preparations with and without the CNS (abdominal ganglion) connected to the gill reveals that the CNS exerts an inhibitory influence on the reflex. Each preparation served as its own control, and sessions were separated by at least 3 hr of rest to eliminate carry-over effects. When the CNS was left connected to the gill and other peripheral structures, the rate of habituation was faster and the response amplitude was lower than after removal, with a stimulus of one water drop per 45 sec. With the CNS intact, it also was more difficult to obtain rapid recovery of responsiveness (dishabituation) by electrical stimulation of the gill than after its removal. We found that severing the branchial nerve resulted in increased amplitude of the withdrawal response, suggesting that the nerve carries to the gill inhibitory effects from the CNS. A cell cluster along the nerve in the gill appears to be involved in this inhibitory activity. The gill without the CNS manifests habituation. We infer from the results, however, that the centrally initiated inhibition serves to rapidly establish the response amplitude appropriate to iterative stimulation.

Blankenship, J. E., Wachtel, H., and Kandel, E. R., 1971, Ionic mechanisms of excitatory, inhibitory, and dual synaptic actions mediated by an identified interneuron in abdominal ganglion of *Aplysia, J. Neurophysiol., 34,* 76–92.

Davis, W. J., and Mpitsos, G. J., 1971, Behavioral choice and habituation in the marine mollusk *Pleurobranchaea californica* MacFarland (Gastropoda, Opesthobranchia), *J. Comp. Physiol., 75,* 207–232.

When stimuli for two different behavior patterns are delivered, feeding is elicited in preference to all other behaviors. The withdrawal response to light habituates as a result of adaptation of the visual response and habituation of the central pathways involving visual input. The sequence of muscular activity of feeding was determined and central nerve cells were located that either excite or inhibit efferent outflow to the feeding apparatus.

Dorsett, D. A., Willows, A. O. D., and Hoyle, G., 1973, The neuronal basis of behavior in *Tritonia.* IV. The central origin of a fixed action pattern demonstrated in the isolated brain, *J. Neurobiol., 4,* 287–300.

Downey, P., and Jahan-Parwar, B., 1972, Cooling as reinforcing stimulus in *Aplysia, Am. Zool., 12,* 507–512.

An experiment was carried out to investigate the role of temperature in the previously reported reinforcing effect of an increase in seawater level in *Aplysia.* In the present experiment, it was found that the reinforcing effect of water level change on rod-pressing behavior in *Aplysia* depends on a decrease in temperature associated with water level change. In order to study modification of rod-pressing behavior produced by contingent increase in water level and decrease in temperature, the rate and latency of rod-press responses in experimental animals were compared with those of yoked control animals exposed to noncontingent water level and temperature change. Higher response rates and shorter response latencies were obtained from experimental over yoked control animals, but only the shorter latencies of experimental animals were attributed to a behavioral change resulting from contingent water level and temperature reinforcement.

Eskin, A., and Strumwasser, F., 1971, Properties of the *Aplysia* visual system: *In vitro* entrainment of the circadian rhythmn; centrifugal regulation, *Fed. Proc., 30,* 666.

Feder, H. M., 1972, Escape responses in marine invertebrates, *Sci. Am., 227,* 92–100.

Limpets, snails, clams, scallops, sea urchins, and other slow-moving sea creatures go into remarkable gyrations when they are approached by a starfish. This lively behavior enables them to deter predation.

Giller, E., Jr., and Schwartz, J. H., 1971, Choline acetyltransferase identified neurons of abdominal ganglion of *Aplysia californica, J. Neurophysiol., 34,* 93–107.

Kupfermann, I., 1974*a*, Feeding behavior in *Aplysia:* A simple system for the study of motivation, *Behav. Biol., 10,* 1–26.

Kupfermann, I., 1974*b*, Dissociation of the appetitive and consummatory phases of feeding behavior in *Aplysia:* A lesion study, *Behav. Biol., 10,* 89–97.

Lukowiak, K., and Jacklet, J. W., 1972, Habituation: A peripheral and central nervous system process in *Aplysia, Fed. Proc., 31,* 405.

The siphon of *Aplysia* exhibits a withdrawal reflex to light and tactile stimuli. This reflex habituates to each stimulus and can be dishabituated by the other stimulus. The time course of habituation and dishabituation are similar in the three types of preparations we have studied: the intact animal, the isolated siphon–mantle–abdominal ganglion, and the isolated siphon–mantle. The abdominal ganglion is not necessary for the siphon–mantle to show habituation and dishabituation. Extracellular recordings from the ctenidial and siphon nerves show that afferent activity is evoked by the reflex and may occur after the siphon contraction has started. Intracellular recording from abdominal ganglion neurons during the reflex or a spontaneous movement show that several types of excitatory and inhibitory synaptic potentials occur, and some decrement occurs during habituation. Kupfermann *et al.* (1970) have emphasized that the abdominal ganglion mediates the related gill reflex habituation, but Peretz (1970) has shown that the isolated gill undergoes habituation. The neuron circuits in the periphery and abdominal ganglion that mediate the siphon withdrawal both undergo habituation and are probably parts of one integrated system.

Maddison, S. E., Hicklin, M. D., Conway, B. D., and Kagan, I. C., 1972, Ionic basis of the photoresponse of *Aplysia* giant neuron: K+ permeability increase, *Science Wash., D.C., 178,* 755–759.

Mpitsos, G., and Davis, R., 1973, Learning: Classical and avoidance conditioning in the mollusk *Pleurobranchaea, Science* (Wash., D.C.), *180,* 317–320.

Naïve specimens of the marine gastropod *Pleurobranchaea* withdraw from tactile stimulation of the oral veil and show feeding responses to food chemicals. Experimental subjects, trained by the pairing of touch (conditioned stimulus) with food chemicals (unconditioned stimulus), soon acquired a classically conditioned feeding response to touch alone. Control subjects that received touch alone or unpaired touch and food chemicals showed significantly fewer feeding responses to touch than did experimentals. Classically conditioned specimens were used for avoidance conditioning. Subjects that received aversive electrical stimulation when they did not withdraw from touch rapidly learned to withdraw rather than feed in response to touch alone. Controls that received touch alone or unpaired touch and shock continued to exhibit the feeding response to touch alone. The learned responses persisted for up to two weeks without reinforcement before extinction and could be demonstrated in the isolated nervous system.

Newby, N. A., 1973, Habituation to light and spontaneous activity in the isolated siphon of *Aplysia:* Pharmacological observations, *Comp. Gen. Pharmacol., 4,* 91–100.

Application of dopamine [(DA)] produces a prolonged slow contracture and stimulates high-frequency brief contractions. DA elevates the amplitude of responses to light. At high doses (10^{-5} and 10^{-4} g per ml) DA retards habituation of the light response; at low doses (10^{-6} g per ml) it speeds habituation. Ergonovine inhibits spontaneous contractions and the light response. Both acetylcholine (ACh) and eserine inhibit spontaneous contractions and the light response. They speed habituation. L-glutamic acid usually produces a contracture similar to that produced by DA, superimposed by high-frequency contractions. This response is quickly followed by inhibition. Light responses are inhibited by L-glutamic acid.

Parmentier, J., 1973, Mapping studies of a gastropod brain, *Brain Res.*, *59*, 201–210.

Peretz, B., and Houreson, D. B., 1973, Central influence on peripherally mediated habituation of an *Aplysia* gill withdrawal response, *J. Comp. Physiol.*, *84*, 1–18.

The parietovisceral ganglion (PVG) modulates gill habituation, which in turn is mediated by a peripheral nerve plexus. A central influence accelerated the rate of habituation and depressed the responsiveness of the gill lobe. The gill plexus is responsible for recovery: PVG influence increased with repeated stimulation.

Peretz, B., Roth, G. I., and Hulette, J., 1971, Acetylcholine involvement in habituated withdrawal responses in *Aplysia* gill, *Fed. Proc.*, *30*, 375.

Habituation has been shown in *Aplysia* gill after removal of CNS; neural organization in the gill appears responsible for habituation; the habituated behavior is a withdrawal movement of one gill lobe (pinnule) in response to water drops falling on it (Peretz, 1970). Rate of habituation is hastened when the gill is continuously infused with acetycholine (ACh), contrasted to infusion of seawater or no infusion. Dishabituation is elicited, as in controls, by electrical stimulation of pinnule and nerve trunks innervating gill. Spontaneous recovery of the behavior is also seen. Histochemistry of the gill shows acetylcholinesterase present along nerve fibers. ACh, presumably accumulating in the gill neural plexus, is suggested as one neurotransmitter mediating habituation.

Peterson, R. P., and Erulkar, S. D., 1973, Parameters of stimulation of RNA synthesis and characterization by hybridization in the molluscan neuron, *Brain Res.*, *60*, 177–190.

Pinsker, H. M., Hining, W. A., Carew, T. J., and Kandel, E. R., 1973, Long-term sensitization of a defensive withdrawal reflex in *Aplysia*, *Science* (Wash., D.C.), *182*, 1039–1042.

When a weak tactile stimulus is applied to the siphon of *Aplysia californica*, the animal withdraws the siphon between the parapodia. This defensive withdrawal reflex can be facilitated (sensitized) if the animal is previously given four days of training, consisting of four brief noxious stimuli each day. The sensitization of this reflex can last for up to three weeks after training and is mediated by the abdominal ganglion, which also mediates long-term habituation. This preparation may provide a system for analyzing the neural mechanism of long-term behavioral modifications of complexity that is intermediate between habituation and associative learning.

Prior, D. J., 1972, A neural correlate of behavioral stimulus intensity discrimination in a mollusk, *J. Exp. Biol.*, *57*, 147–160.

From measurements of membrane electrical properties it was found that the "critical firing level" of the cluster cells is significantly lower than that of the pallial neurons of

Spisula. By way of their differential sensitivities to synaptic input, the stimulus-intensity discrimination could be mediated by cluster cells and pallial neurons directly.

Pusztai, J., and Adam, G., 1974, Learning phenomena in the giant neurons of the snail (*Helix pomatia*), *Comp. Biochem. Physiol.*, *47A*, 165–171.

1. Simple learning phenomena have been investigated in the giant neurons of the esophageal ganglion of the snail (*Helix pomatia*), according to the classical conditioning paradigm. 2. In the course of the experiments, the stimulation of the nervus analis or n. pallialis sinster was paired with an intracellular stimulation, and the formation of conditional hypo- or hyperpolarization has been observed. 3. Elaboration of the conditional response CR required numerous pairings (60–120) and lasted for only a relatively short period of time (15–60 min). 4. It was presumed that the CR is produced by local changes of excitability in the cell membrane. The characteristics of these learned responses (number of pairings, duration of CR, etc.) are similar to Pavlovian summation phenomena.

Rozsa, S. K., and Salanki, J., 1973, Single neuron responses to tactile stimulation of the heart in the snail, *Helix pomatia* L., *J. Comp. Physiol.*, *84*, 267–279.

Stephens, C. L., 1973, Progressive decrements in the activity of *Aplysia* neurones following repeated intracellular stimulation: Implications for habituation, *J. Exp. Biol.*, *58*, 411–421.

Repeated intracellular stimulation of neurons from the isolated abdominal ganglion of *Aplysia californica* produced progressive response decrements with parametric features common to transsynaptic models of habituation. The probability that a constant intracellular pulse of depolarizing current would produce an action potential decreased with repeated stimulation. The response decrements were reversible with prolonged rest. The spontaneous recovery process was long-term in nature. Short, rather than long interstimulus intervals, and a weak, rather than a strong stimulus, produced greater response decrements. These results demonstrate that an individual neurone shows response decrements as a function of repeated stimulation, which suggests that there are at least two processes responsible for the response decrements seen during transsynaptic stimulation: (1) synaptic depression and (2) a depressive process originating in the postsynaptic neuron.

Willows, A. O. D., 1971, Giant brain cells in mollusks, *Sci. Am.*, *224*, 68–75.

Willows, A. O. D., Dorsett, D. A., and Hoyle, G., 1973*a*, The neuronal basis of behavior in *Tritonia*. I. Functional organization of the central nervous system, *J. Neurobiol.*, *4*, 207–237.

Willows, A. O. D., Dorsett, D. A., and Hoyle, G., 1973*b*, The neuronal basis of behavior in *Tritonia*. III. Neuronal mechanism of a fixed action pattern, *J. Neurobiol.*, *4*, 255–285.

Wilson, D. L., and Berry, R. W., 1972, The effect of synaptic stimulation on RNA and protein metabolism in the R2 soma of *Aplysia*, *J. Neurobiol.*, *3*, 369–379.

The Cephalopods (Vol. 3)

Budelmann, B., and Wolff, H. G., 1973, Gravity response from angular acceleration receptors in *Octopus vulgaris*, *J. Comp. Physiol.*, *85*, 283–290.

Carpenter, D. O., 1973, Electrogenic sodium pump and high specific resistance in nerve cell bodies of the squid, *Science* (Wash., D.C.), *179*, 1336–1338.

Horn, G., and Wright, M. J., 1970, Characteristics of transmission failure in the squid stellate ganglion: A study of a simple habituating system, *J. Exp. Biol.*, *52*, 217–231.

The course of transmission failure with repeated stimulation of the squid stellate ganglion for selected stimulus parameters was closely similar to the course of habituation described for interneurons in more complex systems. When a train of shocks was delivered, the synapse potential evoked declined. The time of initiation of the synaptic potential remained constant, but the time of spike initiation became more delayed so that ultimately no spike was generated. Recovery occurred.

Learning in Bacteria and Fungi (Vol. 3)

Diehn, B., 1973, Phototaxis and sensory transduction in *Euglena*, *Science* (Wash., D.C.), *181*, 1009–1015.

INDEX

Ablation (*see also* Lesions)
  octopus nervous system, 54 *ff*
Ammonoidea
  description of subclass, 4
Amplifier junction
  vertical lobe, 87
Anemones
  avoidance of, 13–14
Antedon
  general form, 104
Anthropocentric biases, 147
Aplysia
  habituation mechanisms, 166
  heterosynaptic facilitation, 150–151
Aranea diademata
  functional transfer of response, 149
Argonauta
  brain morphology, 55
Associative learning (*see also* Conditioning and specific paradigms)
  Asterias rubens, 127, 130–132
  echinoderms, 126–133
  octopus, 20 ff
  Pisaster giganteus, 132–133
  Psammechinus, 126–129
  starfish, 133
Asterias
  movement speed, 115
  reaction time, 115
Asterias forbesi
  escape learning, 126–127
  photaxis rhythms, 116
  rhythm reversal attempt, 133
Asterias rubens
  conditioning, 127, 130–132
Asteroidea, 107
Asteroids
  general anatomy, 105
Astropecten
  associative learning, 126–130
Astropecten armatus
  responses to *Pisaster*, 116
Attack inhibition
  octopus vertical lobe, 87
Automation, 141
Avoidance learning
  octopus, 21–22

Bacteria
  chemotactic responses, 181–182
  learning, 179–182
  sizes
    Spirillum, 180
    Spirochaeta, 180
    Thiospirillum, 180
Basal lobes
  effects of lesions, 64, 65
Basiepithelial plexus
  echinoderms, 110–111
Biochemical correlates of learning, 155–156
Blatta orientalis
  functional transfer of reflex, 149
Brain lesions and learning
  octopus, 54 *ff*
Brittle stars
  righting reflex, 122

Cat
  habituation mechanisms, 166–167
Centrostephanus longispinus
  reaction time, 115

Cephalopods, 1 *ff*
  buoyancy, 8
  classification, 3–4
  evolution, 1, 5–6
  laboratory maintenance, 1, 2
  learning, 13 *ff*
  locomotion, 7–9
  morphology, 5–11
  nervous system, 10–13
  reproduction, 6–9
  research subjects, 2, 3
  sense organs, 9–10
  taxonomy 3, 4
Chemotaxis
  as used in bacteria learning studies, 181–182
Circadian clocks
  Chlamydomonas, 179
  Neurospora, 179
Classical conditioning
  critique of invertebrate research, 139–140
Classification
  cephalopods, 3–4
  echinoderms, 106–108
Cockroach
  biochemical correlates of learning, 155–156
  dishabituation, 152
  transfer of training, 149
Coleoidea
  description of subclass, 4
  orders and suborders, 4
Comanthus japonicus
  larval behavior, 134
Comatella
  movement speed, 115
Comparative analyses
  convergent functions, 148–156
  divergent functions, 156–159
Comparative learning mechanisms
  vertebrate–invertebrate comparisons, 168–173
Comparative research, 147 *ff*
Comparative research problems
  species selection, 159–160
  structural considerations, 159
Comparative strategies, 159–168
  polytheticism, 162–168
Conditioning (*see also* Learning and specific learning paradigms)
  Asterias rubens, 127, 130–132
  Astropecten, 126–130
  Eledone, 14, 15, 20
  nervous system units, 153–154
  octopus, 20
  Ophiotrix, 126–129
  Pisaster giganteus, 132–133
  plants, 184–185
  Psammechinus, 126–129
  starfish, 133
Connectionistic vs. statistical theories of learning, 170–172
Control groups
  commentary, 139–140
Convergence
  invertebrate and vertebrate functions, 148–156
Crinoidea, 106
Cristal nerve
  recording, 9
Cucumaria
  general form, 104
Cuttlefish (*see also Sepia*)
  attack response, 23–25
  vertical lobe and learning, 68–70
  vertical lobe and vision, 68–70
  visual discrimination, 32

Delayed reinforcement, 46–47
  vertical lobe, 82–84
Delayed response
  octopus, 43–46
  vertical lobe, 82–84
Dendraster
  responses to *Pisaster brevispina,* 116
Detour experiments
  octopus, 15–17, 58–63
Discrimination
  tactile, 33–35
Discrimination learning
  Astropecten, 126–130
  Eledone, 14, 15
  octopus, 18–19
  Ophiotrix, 126–129
Divergent functions
  vertebrate-invertebrate comparisons, 156–159

Dominant focus
  description, 149–151
  post-tetanic potentiation and
    heterosynaptic facilitation, 150–151
  reflex modification in frog, 150
Drosera
  habituation, 183

Early experience
  epigenetic strategy in learning, 168–169
  invertebrates, 172
Echinarachnius parma
  movement speed, 115
Echinoderms, 103 *ff*
  associative learning, 126–133
  ectoneural nervous system, 113
  electrophysiological research, 134
  entoneural nervous system, 113–114
  escape learning, 122–126
  evolution, 108
  general characteristics, 103–106
  hyponeural nervous system, 113–114
  larval behavior, 134
  learning, 118 *ff*
  movement speed, 114–115
  nervous system, 109–114
  orientation behavior, 115–116
  pedicellariae characteristics, 118
  reflex "republic", 109
  regeneration, 105
  reproduction, 106
  righting reflex, 119–122
  sensory mechanisms, 112–113
  symbiotic activities, 116–117
  taxonomy, 106–108
Echinoidea, 107
Ectoneural nervous system
  echinoderms, 113
Electroconvulsive shock,
  effects on retention, 52–54
Electrophysiological recordings
  octopus, 90–92
Electrophysiology
  echinoderms, 134
Eledone
  conditioning, 20
Eledone moschata
  avoidance of sea anemones, 13, 14

Entoneural nervous system
  echinoderms, 113–114
Epigenetic strategies
  learning, 168–169
Equipotentiality
  Aranea diademata, 149
  Blatta orientalis, 149
  invertebrate and vertebrate similarities,
    148–149
Escape learning
  Asterias forbesi, 126–127
  echinoderms, 122–126
  Ophiura brevispina, 122
  starfish, 122–125
Escherichia coli
  learning, 179, 180
Eupaguras bernhardus
  avoidance by octopus, 13–14
Evolution
  cephalopods, 5, 6
  echinoderms, 108
  learning strategies, 168–173
Experimenter bias
  invertebrate research, 140
Extinction
  Pisaster giganteus, 132–133

Feedback mechanisms
  vertebrate–invertebrate comparisons, 157
Form discrimination
  cue additivity, 32
  octopus, 18–19, 25–33
  theories in octopus, 29–32
Fungi
  learning in *Phycomyces*, 182–183

Generalization
  octopus, 40–43
Generalization and transfer
  vertical lobe, 87
Geotropic responses
  echinoderms, 115–116
Glial cells
  vertebrate–invertebrate comparisons, 158
Gravity
  and form discrimination, 27
Gravity receptors
  cephalopods, 9

Habituation
  cephalopods, 14
  comparative analyses of criteria and mechanisms, 163–167
  criteria, 151–152, 167
  Drosera, 183
  incremental components, 151–153
  mechanisms, 152–153
  Mimosa, 183
  Phycomyces, 182–183
  tactile, 19–20
  transfer in *Limulus,* 149
Heterosynaptic facilitation, 150–151
Holothuria surinamensis
  reaction time, 115
Holothuroidea, 108
Hydra
  habituation mechanisms, 163–164
Hyponeural nervous system
  echinoderms, 113–114

Image-driven behavior, 154
  in spiders, 154
Incremental processes
  and habituation, 151–153
Inferior frontal lobe
  effects of lesions, 66
Innate mechanisms
  and learning, 168
Insect
  nerve cell numbers, 158
Integrative capacities
  invertebrate, 158–159
Interocular transfer
  octopus, 56–62
Intertrial interval
  octopus, 47–48
Invertebrate
  integrative capacities, 158–159
  learning criticisms, 137–138
  models in learning studies, 148
  plasticity (learning) limitations, 159
Invertebrate research
  automation, 141

Larval behavior
  echinoderms, 134

Learning (*see also* Conditioning and specific learning paradigms)
  associative in echinoderms, 126–133
  associative in octopus, 20 *ff*
  Asterias rubens, 127, 130–132
  Astropecten, 126–130
  biochemical correlates, 155–156
  cephalopods, 13 *ff*
  conditioning, 14, 15
  cuttlefish, 23–25
  delayed reinforcement, 46–47
  delayed response, 43–46, 82–84
  detour experiments, 15–17
  discrimination, 14, 15
  discrimination training, 18–19
  echinoderms, 118 *ff*
  escape in echinoderms, 122–126
  Escherichia coli, 179–180
  fungi, 182–183
  generalization, 40–43
  habituation, 14
  intertrial interval, 47–48
  nervous system lesions in octopus, 54 *ff*
  Octopus maya, 22–23
  Ophiotrix, 126–129
  Pisaster giganteus, 132–133
  plants, 182–185
  problem boxes, 17
  proprioception in octopus, 35–36
  Psammechinus, 126–129
  retention, 49–54
  reversal, 36–40
  ribonucleic acid changes, 155
  righting reflex in echinoderms, 119–122
  Salmonella, 181–182
  split-brain effects in octopus, 66–68
  starfish, 133
  synaptic mechanisms, 155–156
  tactile discrimination, 33–35
  tactile system, 63–68
  transfer, 40–43
  vertical lobe in cuttlefish, 68–70
  visual discrimination in octopus, 25–33, 70–77
  visual discrimination and vertical lobe, 70–77
  visual system, 56–62
Learning mechanisms
  epigenesis, 168–169

Learning mechanisms (*cont.*)
innate, 168
suprastructural, 169–173
vertebrate–invertebrate comparisons, 168–173
Lesions
octopus, 54 *ff*
octopus visual system, 56–62
tactile system in octopus, 63–68
vertical lobe of cuttlefish, 68–70
vertical lobe and delayed reinforcement, 82–84
vertical lobe and delayed response, 82–84
vertical lobe function, 86–90
vertical lobe summary, 84–86
vertical lobe and tactile discrimination, 77–82
vertical lobe and visual discrimination, 70–77
Light polarization
discrimination in octopus, 32
Limnaea stagnalis
habituation mechanisms, 166
Limulus
functional transfer of habituation, 149
ganglion organization and habituation, 152–153
habituation mechanisms, 152–153, 164–165
Locust
unit conditioning, 154
Loligo vulgaris
damage, 1
laboratory subject, 1
Lumbricus terrestris
habituation mechanisms, 164

Macrobdella decora
habituation mechanisms, 164
Maze learning
octopus, 22–23
Median inferior frontal lobe
effects of lesions, 65–66
Memory
electroconvulsive shock effects, 52–54
Salmonella, 181–182
short- and long-term, 52–54
transfer, 153
vertical lobe function, 86
Mimosa
conditioning, 184–185
habituation, 183
Morphology
octopus, 5–11
Motivation
vertical lobe, 87
Motor centers
cephalopods, 12, 13
Movement speed
echinoderms, 113–114

Nautiloidea
description of subclass, 4
Nautilus
nervous system, 10–11
phylogeny, 5
Nereis pelagica
habituation mechanisms, 164
Nerve cell
numbers in invertebrate nervous systems, 158
Nerve cell morphology
vertebrate–invertebrate comparisons, 156–157
Nerve cell processes
vertebrate–invertebrate comparisons, 156–157
Nerve centers and learning
vision and touch in octopus, 54–55
Nerve net
basiepithelial plexus, 110–111
Nervous system
cephalopods, 10–13
echinoderms, 109–114
ectoneural in echinoderms, 113
electrophysiology, 90–92
entoneural in echinoderms, 113–114
hyponeural in echinoderms, 113–114
lesions in octopus, 54 *ff*
octopus, 90–92
starfish, 110–111
unit conditioning, 153–154
Neural complexity and capacity, 161

Octopus
arm functions, 6–7
attack, 20
brain morphology, 55

Octopus (*cont.*)
  defense mechanisms, 7
  delayed reinforcement, 46–47, 82–84
  delayed response, 43–46, 82–84
  detour experiments, 15–17
  discrimination learning, 15
  discrimination training, 18–19
  electrophysiological studies, 90–92
  feeding, 7
  general training techniques, 18–19
  generalization, 40–43
  habituation, 14, 19–20
  intertrial interval, 47–48
  learning critique, 142–144
  locomotion, 7, 8
  morphology, 5–11
  mouth, 7
  nervous system, 11–13
  polarized light discrimination, 32
  problem box, 17
  proprioception in learning, 35–36
  retention, 49–54
  retention of tactile discrimination, 49–51
  reversal learning, 36–40
  sensory receptors, 9–10
  short- and long-term memory, 52–54
  tactile discrimination, 33–35, 40
  tactile system, 63–68
  transfer, 40–43
  vertical lobe lesion and visual discrimination, 70–77
  vertical lobe removal summary, 84–86
  vertical lobe and tactile discrimination, 77–82
  visual acuity, 10
  visual discrimination learning, 25–33
  visual system and learning, 56–62
Octopus cyaneus
  conditioning, 20
Octopus maya
  learning, 22–23
Octopus vulgaris
  conditioning, 20
  nervous system, 11–13
  problem box, 17
  suckers, 9, 10
Operant training
  octopus, 20
Ophiotrix
  associative learning, 126–129
Ophiura
  general form, 104
Ophiura brevispina
  escape learning, 122
  righting reflex, 122
Ophiuroidea, 107
Optic lobes
  octopus learning, 56–62
Oreaster nodosus
  learning in righting reflex, 120–121
Orientation
  echinoderms, 115–116
Overtraining reversal effect, 36–38

Paramecium controversy, 140
Partial reinforcement effect
  criticism in earthworms, 141–142
  in earthworms, 137–138
Pedicellariae
  characteristics, 118
Phycomyces
  habituation, 182–183
Phylogeny
  cephalopods, 5, 6
Pisaster
  general form, 104
Pisaster brevispina
  effects on *Dendraster*, 116
Pisaster giganteus
  associative learning, 132–133
Planarian controversy, 140
Plant learning, 183–185
Polarization of brain, 149
Polytheticism,
  in comparative research, 162–168
  and habituation mechanisms, 162–168
Post-tetanic potentiation, 150–151
Predation
  specificity of response in echinoderms, 116
Problem box
  octopus, 17
Procambarus clarkii
  habituation mechanisms, 164–165
  nerve cell numbers, 158
Proprioception
  and learning in octopus, 35–36
Protein changes in learning, 155

Protozoan learning experiments,
as models for bacteria, 180
Psammechinus
associative learning, 126–129
Pycnopodia helianthoides
learning in righting reflex, 121–122
movement speed, 115

Radial symmetry
echinoderms, 104
Rana esculenta
reflex modification, 150
Reaction time
echinoderms, 115
Read-out device
vertical lobe, 87
Reflex "republic"
echinoderms, 109
Regeneration
echinoderms, 105
Reinforcement
delayed, 46–47
delayed and vertical lobe, 82–84
visual discrimination in octopus, 75
Reproduction
echinoderms, 106
octopus, 6, 7
Retention
octopus, 49–54
short- and long-term, 52–54
tactile discrimination in octopus, 49–51, 80–82
Retrograde amnesia
effects on learning in octopus, 53–54
Reversal learning
octopus, 36–40
Rhythm reversal
attempts in *Asterias forbesi,* 133
Ribonucleic acid
changes during learning, 155
memory transfer role, 153
Righting reflex
brittle stars, 122
echinoderms, 119–122
injury, 121
Ophiura brevispina, 122
Oreaster nodosus, 120–121
Pycnopodia helianthoides, 121–122
Rotifers
nerve cell numbers, 158

Salmonella
chemotactic responses, 181–182
Schistocerca gregaria
habituation mechanisms, 165–166
Sense organs
cephalopods, 9–10
Sensitization
classical conditioning, 139–140
and habituation, 152–153
Sensory mechanisms
echinoderms, 112–113
Sensory-motor control
echinoderms, 109
Sepia
attack response, 23–25
brain morphology, 55
visual discrimination, 32
Sepia officinalis
damage, 1
Short-term memory
role of PTP and HSF, 151
vertical lobe, 86
Simple system research,
critique, 162
Slow potentials
as information carriers in learning, 169–173
Special advantage strategy,
in comparative research, 161–162
Spirostomum ambiguum
habituation mechanisms, 163
Split-brain in octopus learning, 56–62, 66–68
Starfish
escape learning, 122–125
nervous system organization, 110–111
righting reflex, 119–122
tube feet functions, 112
Statistical theories of learning, 169–173
Statocysts
cephalopods, 9, 10
Stentor coeruleus
habituation mechanisms, 163
Stichopus panimensis
movement speed, 115
Strongylocentrotas
general form, 104

Structure-function analyses,
in comparative research, 159–161
Subclasses
cephalopods, 4
Suckers
cephalopods, 9, 10
Supraesophageal lobes
lesions and learning in octopus, 56–62
morphology, 56
Suprastructural mechanism in learning,
169–173
Symbiosis
echinoderms, 116–117
Synaptic changes
learning, 155–156

Tactile conditioning
Asterias rubens, 127, 130–132
Tactile discrimination
octopus, 33–35, 40
retention in octopus, 49–51
vertical lobe in octopus, 77–82
Tactile habituation
octopus, 19–20
Tactile system
octopus learning, 63–68
Tactile training
transfer, 40–43
Taxonomy
cephalopods, 3, 4
echinoderms, 106–108
Tetrahymena
conditioning attempts, 180
Transfer
interocular in octopus, 56–62
Transfer of training
octopus, 40–43, 67–68

Tube feet
starfish, 112

Unit conditioning, 153–154
Unit specificities
frog, 154
invertebrates, 155
Urchin
symbiotic relationship with *Aeoliscus strigatus*, 116

Vertical lobe
ablation summary, 84–86
attack inhibitor function, 87
effects of lesions, 64, 65
function in learning, 76–77
learning in cuttlefish, 68–70
lesions and learning, 56–62
motivation role, 87
read-out function, 87
short-term memory, 86
tactile discrimination in octopus, 77–82
visual discrimination learning in octopus,
70–77
Vertical lobe function
summary, 86–90
Vision
cephalopods, 10
vertical lobe in cuttlefish, 68–70
Visual acuity
octopus, 10
Visual discrimination
horizontal–vertical in octopus, 59
nervous system mechanisms, 56–62
octopus, 25–33
Visual system
octopus learning, 56–62